P9-AGQ-258

House Dangerous

House Dangerous

Indoor Pollution in Your Home and Office—and What You Can Do About It!

Ellen J. Greenfield
Foreword by Ralph Nader
Introduction by Arthur B. Sacks, Ph.D.
Director, Institute for Environmental Studies
University of Wisconsin-Madison

Revised and Updated Edition

INTERLINK BOOKS
An imprint of Interlink Publishing Gro
New York

This revised and updated edition was first published in 1991 by

INTERLINK BOOKS
An imprint of Interlink Publishing Group, Inc.
99 Seventh Avenue
Brooklyn, N.Y. 11215

Originally published by Random House, Inc. 1987

Library of Congress Cataloging-in-Publication Data

Greenfield, Ellen J.
House dangerous: indoor pollution in your home and office—and what you can do about it! / Ellen J. Greenfield: foreword by Ralph Nader: introduction by Arthur B. Sacks.
p. cm.
Includes bibliographical references and index.
ISBN 0–940793–64–4 (pbk.)—ISBN 0–940793–68–7
1. Indoor air pollution—Health aspects. 2. Housing and health.
I. Title.
RA577.5.G74 1991
613'.5—dc20 90–5018
CIP

A Blue Cliff Editions Book, Created by Blue Cliff Editions, Inc., New York

Book and cover design: Barry L. Mirenburg
Manufactured in the United States of America

10 9 8 7 6 5 4 3 2 1

To my parents for their love and nurturing,

and to Mark and Jake for their unflagging

support, enthusiasm, affection, and badgering:

This book is dedicated with love.

Acknowledgments

I would like to thank Audrey Thomas for her enthusiastic efforts in guiding me along the torturous routes of research, and Susan Walker and Melinda Corey for their painstaking help in preparing this manuscript.

E.J.G.

Contents

Foreword

by Ralph Nader

Many people view their homes, office buildings, and schools as comfortable sanctuaries from the swirl of airborne pollutants outside, where the smog and dust and stench of industrial pollution seems to reign. Almost all of the publicity, news media attention, and regulatory focus has been on outside pollutants and their smokestack sources. As a result, people have received very little information about their home and workplace air.

Invisible sources of violence

Of course, it is not difficult to smell tobacco smoke, cooking odors, and cleaning fluids. But as Ellen J. Greenfield, the author of *House Dangerous*, walks you through the house, office, and school, we learn that there are hundreds of chemicals, particulates, and other types of exposures that you cannot see, smell, touch, or feel. These substances are the invisible sources of violence to your long-term health, such as cancer-causing, microscopic asbestos fibers, carbon monoxide, polychlorinated biphenyls, and radon. Others, such as formaldehyde and ozone at certain concentrations, may be smelled, but their odors have not yet been identified as danger signals. This means that while their scent may be pungent, they, unlike the smell of smoke heralding a fire, do not drive people to take action. For whatever reasons, all these sources of cumulative damage have largely transcended our sensory alert systems. The old saying that "if it doesn't pinch, it doesn't hurt" does *not* apply for these hazards. So, since we are unable to rely on our senses, we have to rely on our minds, our laws, and our technology to prevent or clean up the air in the interior spaces where most of us spend over 90 percent of our lives.

As is usual with the upward cycle of national arousal over public hazards, a series of seemingly unrelated events take place, which reveals the breadth of a problem and increases our sense of urgency about it. This is what is now happening in the United States with the issue of indoor pollution. In particular, three events are heightening our awareness.

The revolt of the nose

First, there is the growing general revolt of the nose. It goes beyond an annoyance at perfumes and artificial scents: it has to do with protecting our health. All across the country, nonsmokers in ever greater numbers are rebelling against having to inhale the smoke of tobacco users. Where people once considered sidestream smoke simply unsettling, it is now being connected more and more with adverse health consequences.

Second, in 1986, a single event, the surprisingly overdue scientific discovery of dangerous levels of radon gas in over eight million homes in the United States—was a reason for national alarm. This initial alarm will eventually come to be seen as a major galvanizing force in elevating public consciousness. Radon, a naturally occurring radioactive gas, associated with certain geologies in the country, including portions of Montana, Illinois, New York, New Jersey, Pennsylvania,

Colorado, Florida, and Maryland, was found to occur in dangerous levels in some of these areas. Environmental Protection Agency surveys found, for example, that some homes in western Maryland received radon gas concentrations that exposed their inhabitants to the same lung cancer risk as smokers of more than four packs of cigarettes a day. These disturbing findings led the EPA to issue a guideline for radon levels in residences in August 1986, and brought the problem of radon into public discussion. Since that time there has been mounting concern in Congress and elsewhere about the accuracy of radon readings. There is a growing feeling that the government's inspectors leave a lot to be desired.

Third, despite the need for increased EPA activity that events such as the radon discovery demand, the Reagan-Bush regimes have all but closed down the EPA section on indoor pollution, claiming that indoor pollution exposures were not sufficiently important or serious enough to warrant extended study. This withdrawal has honed critics' commentaries, especially regarding the indoor pollution dangers of asbestos, and in so doing, has helped generate public awareness and spark an alert to action.

Benefits of the book

As a consumer call to arms, *House Dangerous* makes a significant leap forward, by clearly outlining the indoor pollutants and their origins in building materials, consumer products, and other sources, and then advising you how you can substantially reduce these contaminants and get your government to do the things you cannot do yourself.

Use in part or in entirety

I am impressed by how determined Greenfield is to make this volume very practical and helpful for you to use—either in part or in its entirety. There are readable chapters on what you can do to lower indoor pollution; the extent to which existing laws can be implemented to curb the sources of pollutants that end up in your home, workplace, or school; a compendium on the major indoor pollutants and their health effects; and easy ways to improve indoor air quality in your kitchen, living room, bedroom-nursery, bathroom, basement, and garage.

If you are the kind of person who needs to have symptoms before you act, *House Dangerous* offers a thorough questionnaire about your health that will allow you to detect current or potential health problems that may be caused by indoor pollution. For example, do you or your work associates have a tendency toward itchy, irritated, or burning eyes, difficulty in breathing, or periodic skin rashes? Some of these symptoms can be linked to contaminants in the air, as studies of the "sick-building syndrome" have found for white collar workers. Further, by using and analyzing the complete indoor pollution checklist that appears in the final chapter, "The Last Word," you can attempt to pinpoint the cause of your symptoms.

Obviously, personal exposure to public hazards can encourage you to become active on a broader plane of reform or correction. It can lead you to know more about the breadth and depth of what needs to be done. To begin with, should manufacturers make public all information about chemical substances that may harm you in aerosol products, pesticides, insulation, building materials, appliances, and other consumer goods? Can, for example, the labeling and package inserts for air freshener be made more explicit, so that you can judge whether that product is worth exposing your lungs to or worth masking other contaminants whose odor may prod you to better ventilate your home or office? Should the federal government be pressured to use existing authorities to impose law and order on corporate coverups or negligence regarding a variety of source contaminants? Are stronger guidelines, mandatory standards, penalties, labeling requirements, or new legislation needed? How can the federal government be induced to do more about the asbestos problem in 30,000 school buildings affecting 15 million students, teachers, and school workers? What can state and local governments do in addition?

The need for reform

Legislation to curb or prevent health problems could also save much anguish and many lives. In this light, should there be a broad, stepped-up monitoring program to test buildings and homes for safety, as fire prevention officials do? (Clearly, better monitoring would have detected the radon problem years ago.) And, with the inevitable profusion of over-the-counter home-detection kits, will government assure consumers of their reliability so that new consumer fraud does not burgeon? These are challenges that you, your local environmental or health group, your labor union, or your parent-teachers' association can and should act upon. The process of effecting change can be complicated and time-consuming, but it is necessary. And it is not without its sources of support. You will find the media quite cooperative in reporting developments that can serve to expand the number of people joining your cause and improve your prospects for success.

In early 1990, an MIT professor and a California physician issued a report on "chemically sensitive" people to the New Jersey Department of Health. The report says it is time for the allergists and clinical ecologists to stop squabbling, and work with public health departments to address this fast-growing affliction.
Consumer and environmental groups are receiving more and more letters from people with low-level exposures to chemicals and gases in the workplace (both factories, labs and "sick" office buildings), in their homes, and outside in what they call "contaminated communities." These people are weary of being referred to psychiatrists as if their complaints are so much hypochondria or delusion. They want more serious research and action.

I am sure that Ellen Greenfield would like to see her book become obsolete for you as soon as possible. Even if this happens only partially, it will be the result of how seriously you take the information and recommendations she has compiled for the defense of your health and your need to breathe purer air indoors. The call to action rests with you.

Introduction

by Arthur B. Sacks, Ph.D.
Director, Institute for Environmental Studies
University of Wisconsin-Madison

Over the past twenty five years, industrialized societies have become aware of short-term and long-term threats to the environment and to public health and safety. Mass media as well as formal and nonformal educational efforts have done much to inform people. Perhaps more than anything else, incidents such as the Chernobyl nuclear disaster, the chemical catastrophe in Bhopal, India, and the Love Canal evacuation have brought environmental issues to the forefront of public concern. Fear of the potential health effects of chemical spills and explosions, toxic exposures in the factory and on the farm, contaminants in drinking water, and carcinogens in foods, pharmaceuticals, and smoking materials has generated intense public response. This concern has created a demand for action to better monitor our environment and regulate the introduction of toxic and hazardous substances, to enforce existing laws, and to remedy current problems.

The threat of indoor pollution

The level of public awareness and involvement in many matters pertaining to environmental health risks is encouraging. Still, one area of potential significance has not yet received adequate public scrutiny. It is increasingly evident that indoor pollution too may pose a serious health hazard. "Indoor" here means the home, white collar offices, schools, and, in general, the supposedly "nonthreatening" world of enclosed personal spaces other than factories.

Over 20 hours per day indoors

Factories have already been acknowledged as posing potential health risks to workers and have, therefore, been subjected to governmental monitoring and regulation. However, in industrialized nations, people spend virtually all of their time—sleeping and waking—in enclosed spaces: homes, offices, stores, restaurants, vehicles. Surveys conducted throughout the industrialized world reveal that most people spend more than 20 hours per day indoors. Many segments of the population—infants and the very young, the elderly, the sick, people who work exclusively in the home—remain in one place, the home, for extended periods each day.

This reality has in itself led to the serious consideration of the potential environmental risks associated with indoor settings. The issue of indoor pollution also became increasingly prominent as a result of the intensive efforts, born out of the energy crisis in the 1970s, to reduce energy costs—using wood or kerosene burners, weatherizing existing dwellings, and building new, tighter structures. Such energy-saving measures do yield lower energy costs, but also introduce new substances to the indoor environment, and reduce air exchange rates. The result is a greater concentration of contaminants and, as *House Dangerous* shows, there are more contaminants in the average modern household and office than many realize.

It is well known that an industrialized society places heavy demands on energy use, that it requires industrial and informational technologies to provide essential goods and services, and that it promotes the development of diverse synthetic susbtances for basic activities such as food production and preservation, manufacturing, and health care. Such advances have been instrumental in increasing life expectancy and in enhancing the quality of life for many. Society has also introduced new elements to modern life, new stresses, new hazards, and new problems along with the new securities and new conveniences. Some contend that we have given up more than we have gained, that we have simply traded old threats for new ones.

Need for more responsibility

These new elements tell us that along with our technological advances comes a need for more responsibility, both to handle the new technology carefully, and to identify and correct problems we have already let get out of hand. We can choose to be prudent. We can decide to investigate these problems and to change our behavior or our laws. Recently, for example, the federal Food and Drug Administration acted to ban the use of sulfites as a preservative in the preparation of certain foods. These chemicals were previously used widely in restaurants to reduce the discoloration and wilting of fruits and vegetables, and were identified as a threat to some half-million people suffering from allergies. Ingestion of sulfites, it was determined, can cause minor illness in some, and can be deadly to others, principally outweighing the benefits gained from their use.

The health effects associated with sulfite use are acute and almost immediately apparent. The health effects of many other substances, however, are more difficult to analyze, and often appear only after long latency periods. In recent years, for example, asbestos has become recognized as a carcinogen. Once revered as a valuable flame retardant for use in homes, schools, factories, and ships, asbestos-containing materials are now being removed from these places with high cost, care, and public interest.

Research is currently underway to identify the types of contaminants present in indoor settings and the potential severity of their short-term and long-term effects on human health. Frequently, the hazards of certain household poisons are well documented, and some communities have instigated local toxic "clean-up" days, providing an easy means to dispose of known toxic substances safely from the home or workplace.

Search more actively for solutions

However, much still remains unknown. We need to awaken to the breadth of the problem and search more actively for its solutions, to invest more in research to understand the dynamics of indoor pollution, to establish the impact of such pollution on human health, to ascertain, where possible, safe and unsafe concentration levels, and to offer mechanisms to control substances and conditions that, by themselves or in combination with others, are deemed harmful.

Much more also needs to be done to educate the public about the scope of this problem. As other instances of environmental hazards have shown, informed citizens can better respond to public policy issues and can make wiser choices about the way they live. To do this, care must be taken to present facts rigorously and objectively.

A first line of defense

House Dangerous will help to draw public attention to this emerging problem and will encourage further government and industrial research. It serves both as a first alert to the multitude of problems and as a first line of defense against them. Those who are well informed about their environment—the immediate world of their households as well as the biosphere as a whole—serve not only themselves, but future generations.

1 Indoor Pollution: The Unsuspected Threat in Your Home

If you are one of millions of safety-conscious Americans, you've probably double-locked your doors, secured your windows, and perhaps even invested in a state-of-the-art home security system. You feel pretty well protected in your home.

And with the high cost of energy for heating and cooling, you may have weatherproofed your home with multiglazed windows, extra insulation in the walls and roof, and caulking or weather-stripping to seal out drafts. You feel pretty safe from the elements, too.

But there is a hidden danger in your home and it is quietly undermining the health and well-being of you and your family. The silent yet growing threat is indoor pollution. You are its captive target.

More pollutants indoors than out

Pollution is most often thought of as an outdoor problem—we can even see it quite graphically in a smog-gray sky or a chemically colored sunset. But researchers note that a greater number of pollutants are generated indoors than outdoors, and pollution in the home, although nearly invisible, is every bit as ominous and, therefore, more insidious. Most of it is virtually odorless, colorless, and detectable only with sensitive scientific equipment. Yet doctors and scientific researchers worldwide are now warning that indoor air pollution may pose an even greater threat than outdoor pollution—primarily because of the greater length of exposure.

Take a breath. Realize that you do this an average of 15 to 40 times each minute (depending on your activity level) for a total of about 21,000 to 58,000 times each day in order to remain alive. The air you breathe needs to be clean and fresh because the respiratory system is a delicate and sensitive entity, efficient in transmitting gases and tiny particulates to the bloodstream. It is so sensitive that allergic persons can react to just one microspore of pollen, and just one tuberculosis bacillus in 50 cubic meters (about 500 cubic feet) of air is sufficient to cause the disease.

Ninety percent of our time indoors

Now think about the amount of time you spend indoors each day. According to a multinational research project done in 1964, employed American men spend an average of 90 percent of their time, or 21.7 hours a day, indoors (6.7 hours at their workplace); 2.9 percent of the day, or 0.7 hours, outdoors; and 1.6 hours in transit. American housewives were found to spend an average of 95 percent of their time, or 22.8 hours a day, indoors (20.5 hours in their own homes); 1.7 percent of their time, or 0.4 hours, outdoors; and 0.8 hours in transit.[1] (See Table A.)

Infants and children, the elderly and the ill, also spend some 95 percent of their time within the confines of the home. They are the first to suffer as victims of residential indoor pollution because of their heightened susceptibility and inherent weaknesses. But the average healthy adult is only somewhat more immune to the effects of pollution.

Table A: Time Spent in Various Locations (average hours per day)

Locations	Employed Men/Women	Housewives
Inside one's home	13.4/15.4	20.5
Just outside one's home	0.2/ 0.0	0.1
At one's workplace	6.7/ 5.2	
In transit	1.6/ 1.3	1.0
In other people's home	0.5/ 0.7	0.8
In places of business	0.7/ 0.9	1.2
In restaurants and bars	0.4/ 0.2	0.1
In all other locations	0.5/ 0.3	0.3

Source: Szalai, A., ed., *The Use of Time. Daily Activities of Urban and Suburban Populations in Twelve Countries* (The Hague: Mouton & Co., 1972), cited in National Research Council, *Indoor Pollutants* (Washington: National Academy Press, 1981), p. V 3.

Looked at in this way, it becomes increasingly clear that the quality of the air we breathe indoors, at home and at work, is critical to our health and well-being, and that increasing concentrations of harmful pollutants in this air constitute a serious threat.

Acute vs. chronic exposures

Exposures fall into two basic categories—acute (high concentrations of a pollutant for a short time) or chronic (low levels of the substance over an extended period). Acute effects of many pollutants are generally apparent: itching or burning eyes or skin, breathing difficulty, choking, constriction of the chest, even death at high concentrations. The adverse effects of chronic exposure to pollutants that remain virtually undetectable to residents, however, are only now unfolding.

Contaminants such as asbestos, formaldehyde, and benzene—found in a wide variety of household products and materials—have been shown to cause illnesses ranging from chronic headaches to cancer. The effects of combustion byproducts such as nitrogen dioxide, carbon monoxide, sulfur dioxide, particulates, and benzo(a)pyrene include impaired lung function, cancer, even death by asphyxiation. Some pollutants that have been measured indoors are known to be mutagenic (causing mutations), embryotoxic (toxic to developing fetuses), allergenic, or damaging to the central nervous system, liver, kidneys, and heart. The National Academy of Sciences estimates that indoor air pollution contributes from $15 billion to $100 billion annually to national health care costs.

Types of contaminants

This growing health hazard is caused in large part by the very things we use to furnish, clean, build, and insulate our homes. A recent study conducted for the U.S. Government Consumer Product Safety Commission (CPSC) discovered that whereas fewer than 10 volatile organic compounds (VOCs) were found to be present in the air outside monitored homes, about 150 VOCs were identified indoors, emanating from plastics, cleaners, polishes, paints, and varnishes, as well as from stoves, heaters, furnishings, insulation materials, and pressed-wood products.[2]

Some popular types of home insulation—most notably urea-formaldehyde foam—have also been identified as dangerous sources of indoor pollution, especially in relatively new homes, mobile homes, and homes that have had additional insulation recently installed. Also, disturbing the existing structure of a house to make additions or improvements can release many harmful substances—such as asbestos—into the home environment.

Energy-efficient homes

But, you may wonder, how did this problem of indoor air pollution suddenly arise? Why did it never appear to be a problem before? There are basically two answers. One has to do with the unfolding realization that many of the medical problems associated with aging—that is, cancer, emphysema, heart disease, senility—most likely stem from long-term environmental stimuli. The other has to do with the recent need to conserve energy, based on the tenfold increase in the cost of fuel over the past decade.

Homes that are designed to save energy are especially effective traps for contaminants. As the ventilation rate—the rate at which stale indoor air is replaced by fresh outdoor air—decreases, the level of pollutants correspondingly increases. The pollutants are literally "trapped" within the shell of the house like air inside the skin of a balloon. And whereas the air in conventional homes is entirely replaced on the average of once an hour, in most energy efficient homes the rate may drop to about once in every five hours—allowing about five times the normal level of pollution to build up.

The increasing use of less conventional energy sources to heat homes also aggravates the problem. Wood- or coal-burning stoves, portable kerosene heaters, and open fireplaces spew an assortment of common and exotic pollutants.

Exposure-related symptoms

Doctors attest to increasing complaints of unexplainable breathing difficulties, bouts of irritability, lethargy, dizziness, vision disturbances, and headaches among patients. Many of these may be attributable to exposure to elevated levels of pollutants. Irritation of mucous membranes leading to increased bronchitis, emphysema, respiratory infections, asthma, and lung impairment can be attributed to such common pollutants as carbon monoxide, nitrogen oxides, sulfur dioxide, asbestos, formaldehyde, aerosol products, bacteria, viruses, and fungi found in higher levels indoors than out.

Sick-building syndrome

In modern office buildings, schools, and public centers, workers have lodged complaints of excessive sleepiness, nausea, dizziness, eye irritation, poor concentration, unpleasant odors, and recurring illness. Employers correspondingly complain of increased absenteeism and decreased production. In cases where these complaints have reached the Environmental Protection Agency or similar state agencies, evidence of what has come to be known as the "sick-building syndrome" has been found: That is, careful monitoring of air quality has uncovered unusually high concentrations of such pollutants as ozone, organic compounds, or formaldehyde, due in great part to severely curtailed ventilation with fresh air.

In many new buildings, and even older buildings where energy-saving measures have been put into effect, the air indoors is almost 100 percent recirculated during the day. In fact, ventilation standards today have been lowered to those of 1830 in the mistaken belief that increased sanitation, high technology, and smoking laws will make up for the lack of fresh air.

Although careful monitoring and legislation of target pollutants such as sulfur dioxide, carbon monoxide, lead, suspended particulates, ozone, and nitrogen dioxide have substantially improved the quality of ambient (outdoor) air, our personal exposure to these and other contaminants has actually increased because of the deteriorating quality of the air within our homes and, often, in our offices, schools, and public facilities.

Limited research done

Still, with all the evidence pointing to a serious and widespread assault on the public health, relatively little outcry had arisen until the mid-1980s. In fact, while well over $25 billion of public and private funds were being spent annually on combatting outdoor air pollution, Congress in 1984 first began appropriating a mere $2 million a year for the Environmental Protection Agency to conduct indoor air research. By 1986, however, evidence of a real threat to public health motivated Congress to enact Title IV of the Superfund Amendments and Reauthorization Act (SARA) to establish a comprehensive research effort to characterize the extent of the indoor air problem and begin to take steps to improve indoor air quality.

It is no doubt disquieting to discover suddenly that the tranquillity of your carefully planned home is under siege from within. Strong emotional and financial ties to our chosen place of residence make all of us less than willing to acknowledge such a pervasive threat. But clearly, ignorance of the problem will not make it go away—it is exactly what contributed to the silent increase in pollution levels in the first place.

IRS and EPA confusion

Today, there exists a growing body of information available to the public on the need for indoor air quality, the sources of pollution and methods of prevention or abatement. Nevertheless, people are just beginning to realize the risks and take action to mitigate them. Contributing to this has been the historical battle between the need to respond to short-term economic conditions and the need to maintain long-term priorities. For example, for years the Internal Revenue Service has offered homeowners substantial tax credits for installing insulation, including urea-formaldehyde foam insulation (UFFI), to cut down on heating fuel consumption. Simultaneously, the Environmental Protection Agency has been attempting to ban the use of UFFI entirely because it is a major source of formaldehyde pollution in the insulated homes.

Homeowners already squeezed by nigh heating costs and high taxes were first encouraged to take advantage of the IRS credit and then, after they learned of the danger of increased formaldehyde levels and perhaps experienced some of the side effects, were forced to spend some $15,000 per house (or about four times the cost of installation) to have the insulation material removed.

Certainly, the escalating costs and decreasing supplies of energy make energy conservation and energy-efficient housing an increasingly important priority. But only complete and accurate information about a product's "pollution potential" will allow you to make sound decisions.

If you carefully select building materials and techniques and monitor their use, you can have a healthier and more fulfilling lifestyle. Ignorance and poor planning, on the other hand, can seriously affect you and your family for years to come.

A tour

The first step toward lowering your own indoor pollution problem is a guided tour through your house or apartment with selected stops at the major sources of pollution.

2 A Walk Through the House

At a busy textile factory in Danville, Virginia, gargantuan machinery swallows up hundreds of thousands of yards of freshly woven fabric. Within this airtight equipment, the fabric is saturated in a bath of chemicals including urea-based formaldehyde, glyoxyls, and other chemical resins. Occupational Health and Safety Administration (OSHA) inspectors regularly and carefully monitor the factory air to make certain it remains uncontaminated by noxious and harmful fumes.

Everyday outgassing

The fabrics that emerge at the far end of the machine are beautifully preshrunk, wrinkle resistant, often stain repellent, and ready to be turned into wash-and-wear clothing or bed linens. But as we wear that clothing and sleep on those sheets, a process known as "outgassing" is sending some of those same toxic chemicals from which plant workers were protected into the air circulating throughout our homes.

True, we are talking here about minuscule amounts—taken alone not nearly enough to be harmful. But added to the hundreds of modern materials we encounter each day—plastics, plywood, particleboard, carpeting, cleaning products—each emitting its own characteristic array of chemicals, we find ourselves bombarded by an increasing number of potentially dangerous substances.

While living with the beauty and convenience of modern materials and appliances that enhance our lives, we are the unwitting hosts to an increasing number of insidious and pervasive pollutants whose presence usually remains undetected but which may be having a profound effect on our health.

How to use the tour

This tour is designed to make it easier for you to spot the often hidden sources of indoor pollution and to help you make decisions on how to cope with them. It is not meant to panic you; only to alert you to a situation that requires your attention. Before you begin the tour, take the simple test beginning on page 69 to help you rate your home's overall pollution potential. Check off the items on the list which apply to your house or apartment and compute your score based on the individual values next to each item. The result will give you a broad idea about your own need to take steps to lower your indoor pollution.

After each room on our tour are a number of "First Steps" you can take to attack the problem areas and lower their impact on your environment. The theory behind these steps is presented in greater detail in subsequent chapters, along with a discussion of your legal recourse from damages suffered because of indoor pollution, the government's position and responsibilities in the matter of indoor pollution, and an extensive Pollution Compendium offering comprehensive information on each major pollutant, its range of adverse health effects, and documented indoor levels.

The Kitchen

Food, family, and friends—the connection goes right to the heart of American society. In fact, the kitchen has become so popular a gathering place that many new homes and apartments have eliminated its walls and placed the kitchen right in the middle of the more traditional living and dining areas, establishing what has come to be known as "the great room." Our tour begins here.

Appliances

According to the "American Housing Survey for the United States in 1987," taken by the Census Bureau for the Department of Housing and Urban Development, there were more than 42.2 million households equipped with gas ranges at that time. That accounts for about 60 percent of American homes. Many cooks swear they would never trade their gas ranges in for the electric variety. But unvented gas stovetops and ovens are responsible for spewing large amounts of carbon dioxide, carbon monoxide (deadly in high concentrations), nitric oxide, and nitrogen dioxide, as well as lesser amounts of formaldehyde, sulfur dioxide, and tiny particulate matter, into the air of your kitchen, to be carried on air currents into the rest of your home.

Gas stoves and dryers

Older gas stoves, equipped with pilot lights, burn continuously, constantly releasing significant amounts of these gases into the indoor environment. The pilots on newer gas stoves apparently release much less nitric oxide and carbon dioxide (owing mainly to more efficient combustion) but about double the amount of carbon monoxide. Stoves equipped with electronic lighting devices instead of pilots offer some relief from continuous emissions, although they remain major polluters when in operation—an average of some three to four hours each day, in many American homes (see Tables A and B). In cases where the stove is additionally used as a supplemental heating source, the hazards are greatly magnified.[1]

Table A: Summary of Pollutant Emission from Gas Stoves

Type of Appliance	Operation	Pollutant Emission Rates (mg/h) NO	NO_2	CO
Older stove with cast iron burners	Pilot lights	6.8	8.2	62.9
	1 burner (high)	250	140	1,031
	3 burners (high)	793	494	3,220
	oven (steady state)	201	161	1,166
New stove with steel burners	Pilot lights	0.5	1.9	84.2
	1 burner (high)	455	277	1,795
	3 burners (high)	1,408	699	3,213
	Oven (steady state)	171	111	3,564

(No pans were in place)

Source: NRC, *Indoor Pollutants* based on Cote et al., "A Study of Indoor Air Quality"

Especially in winter, when houses tend to be shut tightly, the level of the gases released by the stove soon average out over the entire house or apartment, including the bedrooms, where people spend a large portion of their day.

Exhaust fumes

Although much less common today, some homes are also equipped with gas-powered clothes dryers, which, unless vented to the outdoors, also expel the same combustion pollutants into the home. Although there are so-called energy-saving devices on the market supposedly designed to clean and recapture the residual heat in the air from the exhaust, these usually consist of little more than a lint filter in a frame made to fit over the dryer's exhaust hose. The gaseous portion of the exhaust fumes is still expelled indoors, along with a sizable portion of the smaller, and therefore more respirable, particles. The problem of respirable particle emission is also inherent in electric clothes dryers that are not vented outdoors.

Table B: Summary of Gas Stove Emissions

Pollutant	Oven[a] Emission[c]	Top Burner[b] Emission[c]
Gases:		
CO	950 (650-1,600)	890 (720-1,090)
CO_2	200,000 (195,000-201,000)	205,000 (196,000-217,000)
NO	29 (14-50)	31 (21-47)
NO_2	62 (44-74)	85 (69-100)
SO_2	0.8 (0.9-1.0)	0.8 (0.6-0.9)
HCN	1.8 (1.6-2.3)	0.07
HCHO	11.4 (9.9-14.2)	5.2 (2.0-12.0)
Particles:		
Carbon	0.13 (0.05-0.24)	0.90 (0.86-0.96)
Sulfate	0.01	0.05 (0.01-0.08)

[a]Oven operated for 1 hour at 350°F.
[b]Operated with water-filled cooking pots.
[c]Given in micrograms/kcal. Ranges in parentheses.

Source: NRC, *Indoor Pollution*

It has been well documented that indoor nitrogen dioxide levels often exceed ambient (outdoor) levels and sometimes remain as much as five times higher, where gas-burning appliances are used and room ventilation is limited. The short-term (five-minute to one-hour) concentrations have been found to be as much as ten times higher than national air quality standards allow outdoors. Carbon monoxide concentrations have also been found to exceed health standards in gas-fueled homes.[2,3]

Researchers at California's Lawrence Berkeley Laboratory studied the emissions of gas ovens operating at 350 degrees Fahrenheit and found that in a small, poorly ventilated kitchen, the air could soon contain as much carbon monoxide and nitrogen dioxide per cubic foot as the air over Los Angeles on a bad day.[4]

Respiratory symptoms

The implications for our health, and particularly for the health of our children, seem clear. Indeed, two British studies by R.J. Melia et al., that each examined some 5,000 children age 5 to 11 years old noted a higher incidence of respiratory symptoms and disease in those living in homes with gas cook-stoves than those in electric-cooking homes. The study reported that this was independent of age, sex, social class, number of cigarette smokers in the home, and geographic latitude (although the correlation only appeared for urban areas). In one of the two studies, it appeared that the effect declined as the children grew older. Another British study by Florey et al. that examined more than 800 children six to seven years old, also found greater prevalence of respiratory illness among children from gas-cooking homes, which coincided with increased levels of nitrogen dioxide in the children's bedrooms.[5]

An American study by Frank Speizer of Harvard University, involving some 8,000 children six to ten years of age, linked increased respiratory disease before the age of two, and impaired lung functions, with the presence of gas stoves.[6] Several other studies have had similar results. In most cases, the adverse health effects are presumed to be caused by the elevated levels of nitrogen dioxide.[7] Dr. John Spengler, of Harvard University's School of Public Health, emphasized that any changes in the respiratory system that take place in childhood may have increased significance as a person ages.

Despite this evidence, there is no clear consensus on the health effects of gas stoves, because other studies have failed to note a correlation between acute respiratory illness or lower levels of pulmonary function and gas stove use.

Cleaning products

Appliances are not the only sources of pollution in the kitchen. Household cleaning products—oven cleaners, air fresheners and disinfectants, furniture and floor polishes, fabric cleaners and stain repellents, drain cleaners and bleaches—contain an astonishing array of organic compounds, solvents, and toxic chemicals such as ammonia and chlorine that enter the indoor environment. In many cases, the ingredients are not displayed or are trade secrets, leaving consumers completely unaware of just what they are introducing into their homes (see Table C).

Table C: Typical Ingredients of Selected Household Cleaners and Solvents

Note that virtually all of the listed ingredients can be classified as toxic or producing adverse health effects if ingested or inhaled in high concentrations or over prolonged periods. In general, however, they are currently regarded as safe for their intended use in a well-ventilated area. Decreased ventilation could well have health implications.

Product Type[1]	Common Ingredients[2]
Window cleaner	Ammonia; Ammonium hydroxide; Ethylene glycol; Isopropyl alcohol; Sodium nitrate
Drain cleaner	Petroleum distillates; Sodium hydroxide (lye); Trichloroethane
Spot remover	Ammonia hydroxide; Benzene; Carbon tetrachloride; Methanol; Naphthalene; Perchloroethylene; Sodium hypochlorite; Toluene; Trichloroethane; Trichloroethylene
Prewash treatment	Perchloroethylene; Petroleum distillates
Polishing cleanser	Chlorine
Oven cleaner	Hydroxethyl cellulose; Polyoxyethylene fatty ethers; Sodium hydroxide (lye)
Furniture polish	Dinitrobenzene; Petroleum distillates; Silicone; Trichloroethane; Wax morpholine
Tile cleaner	Tetrasodiumethylenediamine
Disinfectants (Lysol)	Carbolic acid; O-phenylphenol; N-alkyl-N-ethyl morpholinium Ethyl sulfates
Disinfectant sprays	Triisopropanolamine morpholine
Bleach (Clorox)	4-Chloro-2-cyclopentylphenol; Diethanolamide-lauric acid amide
Air fresheners	Cresol; Ethanol; Propylene glycol morpholine
Paint removers, thinners	Acetone; Benzene; Ethyl ether; Methanol; Methylene chloride; Naphthalene; Toluene; Turpentine; Xylene; Xylol

[1]For less toxic alternatives to these products, see Product Substitution Table in chapter 4.

[2]These are "typical" components. Your chosen brand may have more or fewer ingredients than those listed here.

Among the compounds found within the home are many mucous-membrane irritants, such as ammonia, turpentine, toluene, sodium hydroxide (lye), perchloroethylene, naphthalene, acetone, tridecane, pentadecane, and hexadecane. Benzene and chlordane are human carcinogens (and methylene chloride is highly suspected of the same), while other chemicals found indoors, such as xylene, diethylphthalate, dodecane, tetradecane, methylene chloride, perchlorethylene, dichloroethane, and trichloropropane are mutagenic or embryotoxic (causing birth defects or fetal death). Methylene chloride also causes increased carboxyhemoglobin levels in the blood (decreasing the blood's ability to carry life-giving oxygen to the cells) and has been reported to cause brain damage, and perchloroethylene fumes can produce headaches and have a narcotic effect. Although not all the compounds named in Table C cause adverse health effects at commonly encountered indoor concentrations, the list of those implicated in health problems seems to grow almost monthly, as new research surfaces.

Propellents

In addition, many of these products are dispensed by means of aerosol cans, which release a significant amount of propellent along with the product at hand. In an average application, the following amounts of propellent are found to be released: 6–12 grams with air freshener; 8 grams with disinfectant; 8 grams with furniture polish; 4–8 grams with dusting spray; and 20–25 grams with oven cleaners. It has been estimated that an average suburban household generates some 15 pounds of propellent per year.[8]

Because of the well-publicized threat to the ozone layer of our atmosphere, fluorocarbon propellents, which were widely used to dispense household products in the 1960s, have been replaced by flammable hydrocarbons, such as propane, nitrous oxide, and methylene chloride. These two latter propellents have an anesthetic (deadening) effect on the central nervous system, and methylene chloride, when inhaled, can convert to toxic carbon monoxide. In addition, this chemical causes cancer in laboratory animals.

It has been estimated that more than 40 aerosol spray products can be found in the average American home. They present a double problem because not only does their propellent gas remain airborne permanently but they also scatter the product they are meant to dispense into the surrounding air in the form of a fine and easily breathable mist.

Paper products

Some other kitchen polluters are less pervasive, but still present. These include the wide range of paper products—waxed paper, grocery bags, paper plates and cups, and paper towels among them—that contain formaldehyde, introduced during the manufacturing process as an adhesive to increase the wet strength of the paper. This formaldehyde, too, is released in your kitchen, although its relative impact on the quality of your environment is not nearly as significant as that of particleboard, plywood, and paneling.

Cookware

Even ordinary pots and pans may pose a hazard to your health. Although stainless steel and cast iron cookware are commonly accepted as superior in terms of heat distribution and sturdiness, aluminum pots and pans are still in common use and are readily and cheaply available. Cooking and stirring may release some of this aluminum into your food, and ingestion of this can cause above-normal levels of aluminum to build up in your body. Abnormally high levels of aluminum have been found in many victims of Alzheimer's disease, sometimes known as premature senility, although no positive cause and effect relationship has yet been discovered.

Microwave ovens

Microwave ovens, which in their early days emitted enough stray radiation to cause concern, especially for those who wore cardiac pacemakers that could be affected by the frequency of the microwaves, are now considered safe by most sources, including the Consumer Product Safety Commission. Stringent emission control regulations for these ovens and for color television sets are said to have eliminated virtually all hazard connected with their proper use. The exception, at least with microwaves, is in the case of a defective door, which can then leak microwaves in concentrations that are potentially damaging to human tissue.

Microwave radiation differs from nuclear radiation, a much more powerful sort of electromagnetic energy that can create ionized molecules in the material it penetrates, thus changing the material's inherent structure or the key to its reproduction. The most recent tests on microwave ovens by *Consumer Reports* magazine found that leakage did not exceed 0.5 milliwatts per square centimeter (0.5 mW/cm^2), and that it declined precipitously as the distance between person and oven increased. At about four feet (which the magazine considered a reasonable distance from the appliance for someone working in the kitchen), radiation levels averaged roughly 0.01 mW/cm^2. That is about 100 times less than the maximum leakage level permitted by federal standards for new ovens at a two-inch distance. The magazine also noted that extensive research indicates that even extended microwave exposure of 0.1 to 1 mW/cm^2 has no apparent harmful effect on humans. Occupational exposures at that level are considered safe even over the course of an entire working day.

Safety

For extra safety, the magazine cautioned, don't spend long periods peering into the oven door watching your food cook, and if you're pregnant, don't linger near the operating oven. In order to protect the door seals, do not use swing-down doors as shelves.[9]

First Steps

■ **If you have a gas stove and oven:**

Make sure burners are properly adjusted (flame should burn blue);

Install a range hood vented to the outdoors far from windows and other air intake sites;

If venting outdoors is impossible, open a nearby

window when stove or oven is in use. This is even more effective if an exhaust fan is placed in the window;

Never use oven as a supplemental heating source.

- **Purchase only pilotless gas appliances.**

- **Do not use gas clothing dryers unless they are vented outside and away from air intake sites.**

- **Whenever possible, replace gas appliances with electric appliances.**

- **If you have a microwave oven, preserve the integrity of door seals by never using fold-down door as a shelf.**

- **Avoid aerosol spray cans whenever possible.**

- **Substitute low-polluting cleaning products and techniques whenever possible. (See Product Substitution Table in chapter 4.)**

- **Keep only as many paper bags and other paper products as you know you will use.**

- **Substitute stainless steel, ceramic, or cast iron cookware for aluminum pots and pans.**

- **Use an efficient air cleaner to minimize unavoidable pollutants. (More on these in chapter 4.)**

The Living Room

Fireplace and heating stoves

Even today, in our advanced age of comfortable central heating, the fireplace remains a potent symbol of family life. But the fireplace that brightens our winter evenings also sends a variety of undesirable pollutants into our homes.

All substances, including wood and paper, give off carbon monoxide, nitrogen oxides, particulate matter, and organic chemicals when burned. And when the fireplace or wood stove is being used, some of these pollutants enter the room, even as the rest goes up the chimney or flue. Cracks in a stovepipe, downdrafts, or spillage of wood from the fireplace can exacerbate this condition.

Woodsmoke

Although the range of pollutants emitted in woodsmoke have been carefully studied (see Table D), relatively little is known about the exact level of these emissions and their impact on indoor air quality. Nevertheless, there is a great deal of concern about the organic compounds and particulates formed during residential woodburning, based on the large number of carcinogens involved and the fact that almost all the particulates are within the respirable-size range.

Scientists also stress that the large number of pollutants in woodsmoke is especially disturbing because of the likelihood that these chemicals may act in conjunction with one another to add up to a greater health threat than any of them would alone.

Table D: Air Pollutants in Woodsmoke

Pollutant	Carcinogenic Activity
Carbon monoxide	
Nitrogen dioxide	
Sulfur dioxide	
Acenaphthylene	
Fluorene	
Anthracene/phenanthrene	
Phenols[1]	*
Fluoranthene	*
Pyrene	*
Benz(a)anthracene	+
Chrysene	+/−
Benzofluoranthenes	+ +
Benzo(a)pyrene	+ + +
Indeno pyrene	+
Benzo(ghi)perylene	
Dibenzanthracenes	+ + +
Ancenapthene	
Ethyl benzene	
Phenathrene	
Dimethylbenzanthracene	+ + + +
Benzo(c)phenanthrene	+ + +
Methylcholanthene	+ + + +
Dibenzopyrenes	+ + +
Dibenzocarbozoles	+/− to + + +
Formaldehyde[1]	
Propionaldehyde[1]	
Acetaldehyde[1]	
Isobutyraldehyde[1]	
Cresols[1]	
Catechol	*

[1]Cilia toxic and mucous coagulating agent

+/− Uncertain or weak carcinogen; + carcinogen; + +, + + +, + + + + strong carcinogen; * initiating or cancer-promoting agents and co-carcinogenic compound

Modified from: Spengler, J.D. and Cohen, M.A., "Emissions from Indoor Combustion Sources," *Indoor Air and Human Health,* Gammage, R.B. and Kaye, S.V., eds.; and Calle, E.V. and Zeighami, Elaine A., "Health Risk Assessment of Residential Wood Combustion," *Indoor Air Quality,* Walsh, P.J., Dudney, C.S., and Copenhaver, E.D.

Of particular interest to researchers is the chemical benzo(a)pyrene, which is also found in cigarette smoke and is considered a strong carcinogen. A field-monitoring program of wood-burning homes in Boston found that use of the wood stove increased the indoor benzo(a)pyrene concentration five times.[10]

It should also be noted that burning coated stock paper or colored print paper such as magazines or newspaper comics introduces a trace of arsenic vapor and the volatile organic chemicals present in their inks or coating, both of which could cause acute respiratory symptoms in poorly ventilated rooms. Synthetic fireplace logs, basically made of wood pulp and chips bound with highly flammable resins and alcohols, are also imbued with chemicals that give off bright colors when burned. Traces of these chemicals, too, persist in the air we breathe, adding to what scientists call our body burden of toxic or potentially carcinogenic compounds.

Coal stoves

Coal stoves add their own special breed of pollution in the form of high levels of sulfur dioxide, the same chemical responsible for the deadly smogs that blighted many parts of England for more than 500 years. Today, sulfur dioxide, as a by-product of coal burning, is best known for its connection with the acid rain that is slowly killing the lakes and forests of the Northeast.

Wood or coal stoves also require particular attention in that their combustion by-products are more completely enclosed within the confines of the home. In a fireplace, as much as 90 percent of the fire's heat and exhaust gases are lost up the chimney. In a stove, however, the air supply and rate of burning is controlled, so that as much as 60 percent of the heat output is delivered indoors. The flue damper, through which the exhaust gases pass on their way up the stovepipe, can be finely regulated, as can the air inlet, to control the rate of burn. If this is done incorrectly, or if the stove has been improperly installed or maintained, the level of combustion products entering the room can quickly exceed the limits of safety.

Indoor emissions reach their peak whenever the door to the stove is opened for stoking or adding fuel, causing a back draft. To avoid this, the damper should be opened before the stove door to increase the draft up the flue. Loose stovepipe joints and leaky door gaskets can also allow significant emissions to escape indoors. A chimney that is too short can allow smoke from outside to re-enter at windows, doors, or air vents.

Formaldehyde-containing materials

That ubiquitous pollutant, formaldehyde, is also being exuded in your living room, by such common items as drapes, carpet backing, decorative paneling, partition walls, furniture made using industrial particleboard or medium density fiberboard (as most furniture is these days), upholstery, and carpet shampoos. Of course, many of the same items discussed here as containing formaldehyde appear elsewhere throughout your home and should be considered in evaluating your own indoor air quality.

Formaldehyde is a mucous membrane irritant affecting skin, eyes, nose, and upper respiratory system. In sensitive individuals, even extremely low concentrations may trigger a strong reaction. High concentrations may be responsible for causing asthma and similar lower airway and pulmonary effects. Some researchers believe formaldehyde to be carcinogenic, although debate rages over this. At very high levels, formaldehyde is lethal to humans.

Plywood, floor coverings

In 1981 formaldehyde was twenty-fourth among bulk chemicals produced in the U.S. (about eight billion pounds per year), and because of its superior bonding properties and low cost, it is used in a wide variety of products. Plywood, for example, is made of thin sheets of wood glued together with urea-formaldehyde (UF) resin, and particleboard is made by saturating wood shavings with UF resin and pressing the resulting mixture, usually at high temperature, into the desired form. The forest-products industry uses some 500,000 tons of formaldehyde resins annually in the manufacture of these two products.

Formaldehyde polymers are also used extensively in manufacturing floor coverings and carpet backing, and UF resins appear in home textiles to impart water repellency, wrinkle-resistance, fire retardance, and stiffness. Formaldehyde also acts as a disinfectant, and as such appears in some carpet shampoos.

The individual emission rate of each formaldehyde-containing item in your home or apartment is likely to be small (except in the case of partition walls and some furniture), but even minimal outgassing by each product may result in a significant increase of formaldehyde to your indoor environment when the ventilation rate is low. Because of this, indoor formaldehyde levels are found virtually always to exceed outdoor levels.

Hot and humid conditions usually cause formaldehyde to outgas at a greater rate. Aging of the product diminishes its emission rate, although in some cases it may be several years before a significant decrease occurs.

Carpets

Carpet shampoos deserve a special mention here, as they have been cited in many cases as respiratory irritants. Daycare centers and schools with carpeted classrooms have often been the site of mass outbreaks of upper respiratory infections and other flu-like symptoms after carpet cleaning, as have offices and motels. Although the carpet shampoo was implicated, the exact ingredients responsible have yet to be pinned down. Some researchers, however, believe the culprit to be the powdery residue of the detergent sodium decyl sulfate (also known as sodium lauryl sulfate), an anionic detergent used widely in laundry detergents, cosmetics, shampoos, toothpastes, and food. Long known to be an eye and skin irritant, it is only now being tested for its respiratory irritancy.[11]

First Steps

■ **Install an air intake duct at bottom of fireplace to allow the fire to draw outside air for combustion.**

■ **Keep burning wood well inside the fireplace.**

■ **Avoid burning synthetic fireplace logs, coated stock paper, color-printed paper, and newspaper.**

■ **Install glass fireplace doors to lessen particulate emission.**

■ **Avoid burning coal.**

■ **Have wood stoves, flues, and chimneys properly installed, allowing ample room for smoke to escape far from air intake sites.**

Around the house

■ **When shopping for a wood stove, look for one with a secondary combustion chamber and/or catalytic converter to reduce emission of hydrocarbons.**

■ **Periodically check gaskets and joints of stove to make sure they are tight.**

■ **Try to buy solid wood furniture rather than that made mainly of composite materials such as plywood and particleboard.**

■ **Avoid buying carpets or carpet pads that smell overwhelmingly of formaldehyde. If after**

sniffing a product you have a strong reaction or detect a strong pungent formaldehyde smell choose another product.

- **Hardwood floors and small area rugs hold less dust than wall-to-wall carpeting and are also less likely to be strong sources of formaldehyde and mold spores.**

- **Have carpets and rugs professionally steam-cleaned.**

- **Especially avoid home carpet-cleaning sprays.**

- **Use an efficient air cleaner. (More on this in chapter 4.)**

Bedrooms and Nursery

The person who lives the average seventy-two-year lifespan will have spent a full twenty-four years of that time asleep. During the first year of life, the newborn will have spent a total of some twenty-eight weeks asleep (although it may not seem that way to new parents). In addition, much of the infant's waking time is generally spent within the confines of the nursery. The elderly and the ill also spend a great deal of their lives in the bedroom. This makes the safety and quality of the bedroom and nursery environments of the greatest concern to us all. Yet many of the things we do to increase our comfort in the bedroom may actually jeopardize the environmental quality there.

Formaldehyde

As noted at the very beginning of this chapter, formaldehyde can become an uninvited overnight guest in your bedroom as it slowly outgasses from your new bed linens and the wrinkle-resistant cotton and cotton-blend clothing in your closet. Although the rate of formaldehyde emanation from these sources is small compared with such sources as plywood and particleboard, it does add to the overall pollution burden of your home. Fortunately, in the case of textiles, reducing the extent of the problem is simple: Laundering newly purchased clothing and bed linens before using or storing will virtually eliminate the excess formaldehyde and sizing chemicals (used to impart the desired "hand" to fabrics).

Kerosene space heaters

In 1982, as expensive fuel oil made home heating costs soar, more than five million portable kerosene space heaters were purchased in this country. Household thermostats were turned way down to conserve precious fuel, but individual comfort was maintained by turning up the portable heater in one or two rooms, most often the bedrooms or family room.

The heaters were reputed to be safe from risk of fire or burns when used as directed, but they were never free of harmful pollutants delivered directly into the closed room along with the heat. The most lethal of these is carbon monoxide, but others include carbon dioxide, nitrogen dioxide, sulfur dioxide, and formaldehyde. (See Table E.)

Table E: Pollutant Emissions from Unvented Gas-Fired Space Heaters

Operation	Pollutant Emission Rates (mg/h)		
	NO	NO_2	CO
Low flame, steady heat	214	130	1,770
High flame	837	272	1,982

Source: NRC, *Indoor Pollutants*, from Cote et al. "A Study of Indoor Air Quality"

Consumers Union, publisher of *Consumer Reports*, calculated the concentration of four gases produced by the heaters—carbon monoxide, carbon dioxide, nitrogen dioxide, and sulfur dioxide—in 10 x 12 x 8 foot rooms with normal ventilation. The levels of each gas were high enough to be a serious health hazard to high-risk groups, including pregnant women, asthmatics, people with cardiovascular disease, children, and the elderly. In fact, the levels calculated for some of these gases would pose a threat even to healthy adults.[12]

Other studies corroborate these results. In tests run by Lawrence Berkeley Laboratory, carbon dioxide levels reached 10,000 parts per million (ppm), twice the eight-hour occupational health standard, after one hour of heater operation. Nitrogen dioxide levels were up to seven times higher than the California one-hour standard of 0.25 ppm and carbon monoxide concentrations were 1.4 times the eight-hour EPA standard of 9 ppm. An improperly adjusted wick could boost this to ten times the standard.[13]

Sulfur dioxide, nitrogen dioxide

Yale University researchers found that kerosene heaters "can result in concentrations of sulfur dioxide and nitrogen dioxide in excess of the relevant ambient air quality standards," and that carbon dioxide levels could be higher than allowable OSHA standards.[14] In essence, federal safety and health regulations would prohibit a coal miner from working in such conditions as a kerosene heater might produce in your bedroom or nursery.

Nevertheless, there are some twenty brands of kerosene heaters still on the market and more than ten million kerosene heaters in use throughout this country. Several states, including Colorado, Oregon, and Washington, have banned the sale of unvented kerosene heaters, and in California, retailers must warn customers not to use them in their homes. Still, that is where most kerosene heaters wind up.

In addition to the more popular kerosene heaters, gas-fired space heaters are also available. Although the exact rate of emissions of various gases are somewhat different for these, the basic danger remains the same.

Humidifiers and air conditioners

Although a nice, warm room seems cozy and inviting during the frosty days and nights of winter, heated air is dry air, and hours of breathing in such an arid climate can dry up the protective coatings of the delicate mucous membranes in the nose and throat. The result is often greater susceptibility to cold and flu viruses. As this is becoming more widely recognized, more people are using humidifiers to replace some of the water vapor lost in the heated household environment. But here we are caught between a rock and a hard place, because the very devices we use to replace the lost water vapor in the air often harbor and disseminate a plethora of bacteria, viruses, and spores to make us ill.

Cool-mist humidifiers

Cool-mist humidifiers, which are often recommended by pediatricians for use in nurseries to avoid the stresses that artificially heated, dry air can place on young lungs, can be breeders and dispensers of these live pollutants. Instead of relieving the symptoms of colds and sore throats, humidifiers that are not regularly and adequately cleaned and disinfected can actually be spreading the causes of infection and allergic reaction.

Fungi, molds, amoebae

A complex mix of fungi, molds, amoebae, and bacteria find the water reservoirs of humidifiers a perfect environment for growth. Some researchers believe that bacterial contamination of these units approaches 100 percent. Stringent cleaning and disinfection of the water reservoir and mist mechanism can cut down on microbial growth, but only temporarily. Allergic individuals will react to the mold spores and fungi released from the humidifier and spread in a fine aerosol into the room.

I experienced this first-hand one year, when my then year-old son suffered from deep congestion, breathing difficulties, and a runny nose from November to May, while I diligently ran his cool-mist humidifier, night and day, trying to relieve his discomfort. With the advent of the balmier days of spring, I stopped using the machine, and his ailments abruptly ceased. The following winter I avoided the humidifier and his colds were all short-lived and milder. Although I can claim no scientific proof of a direct link between the humidifier and my son's incessant illness, the evidence points in that direction.

In office buildings and factories, a condition known as "humidifier lung," characterized mainly by fever, chills, headaches, chest tightness, and breathing difficulty, is usually traced to the building-wide humidifier that adds moisture to incoming air after it has been heated.

Hypersensitivity pneumonitis and Legionnaire's disease are also cultured in and transmitted via humidifiers. These illnesses are, of course, much more serious, but also are far less common.

Ultrasonic humidifiers

According to research done by Consumers Union, ultrasonic humidifiers disseminate far fewer microbes than the more common and less expensive mechanical kind, although the reason for this is not clear. The organization tested some 18 different ultrasonic humidifiers and four conventional cool-mist models. After running them almost continuously for three weeks using tap water, they allowed the humidifiers to sit idle for a week. By that time mold had grown in the water of all the tested models. However, when each model was then run in a disinfected room along with petri dishes filled with media designed to grow molds and bacteria, only the conventional humidifiers triggered any significant microbial growth. The researchers theorized that the ultrasonic vibrations somehow destroyed the microbes that grew in the standing water, perhaps by breaking them apart.[15]

Despite the failure of ultrasonic humidifiers to produce viable microorganisms, research indicates that people with asthmatic or other highly allergic conditions could still react to the pieces of fungi or other microbes spewed into the air. Because of this, ultrasonic humidifiers, too, should be cleaned thoroughly on a regular basis according to the manufacturers' instructions.

The Consumer Product Safety Commission (CPSC) released a "safety alert" in December 1988 regarding the health hazards of breathing microbe-contaminated mist from conventional humidifiers, but adding the additional caveat that the mist from ultrasonic humidifiers harbored its own hazard in the form of respirable particulates. The particles in this mist are comprised of the impurities in tap water, which may contain lead, aluminum, asbestos, or dissolved organic gases. In one experiment, the fine particle concentration was found to be higher than the NIOSH limit for respirable particulates and nearly 50 times the EPA's standard for ambient air. The CPSC therefore, recommends using distilled water with ultrasonic humidifiers.

Air conditioners, too, provide ideal conditions for breeding and dispensing microbes. The coils remain cool and wet and when contaminated, microorganisms are readily blown from these surfaces into the room. Air conditioners, however, seem to be less efficient breeders of these microbes in the home than are humidifiers.

First Steps

■ **Wash all bed linens and clothing before using or storing.**

■ **Use only outside-vented gas or kerosene heaters or electric heaters to supplement home heating system.**

■ **Clean all humidifiers daily with a strong solution of vinegar and hot water.**

■ **If you do use a cool-mist humidifier, start it with very hot water, which contains fewer microorganisms, and never allow leftover water to stand for long periods.**

■ **Substitute ultrasonic humidifiers for cool-mist models.**

■ **Service and repair air conditioners when needed, and clean coils occasionally, checking for visible microbial growth.**

■ **Use an air cleaner to keep bedroom air particularly pollution-free. This can be especially important for asthma or allergy sufferers. (More on these in chapter 4.)**

The Bathroom

A breeding ground

The problem of mold, mildew, and bacteria is also common to the bathroom, where standing water in the toilet makes an excellent breeding ground, and flushing provides an efficient means of microbial dispersal. Stall showers that remain damp long after use are also subject to significant mold growth, and spores easily travel the air currents from this room to the rest of your house. In addition, showering, toilet flushing, and other forms of water agitation help to release any radon gas contained in your water supply. (See Radon under Building Superstructure in this chapter.)

Aerosol products

Bathrooms are also home to many aerosol products. Hair spray, spray deodorants, perfumes, shaving creams, air fresheners, even disinfectants and tub and tile cleaning preparations, are often dispensed by spray cans powered by the same volatile hydrocarbons discussed under Cleaning Preparations. For each average application of deodorant, for example, a gram of propellent is released, and for shaving cream, a half gram is released. Hair spray releases five to seven grams of propellent for each use, centered directly around the nose and face. One common solvent, methylene chloride, had until recently been found in about 25 percent of all hair sprays. In 1989, the Food and Drug Administration banned its use in cosmetic products because it had proven to be carcinogenic in laboratory animals. Methylene chloride had sometimes been indicated on the label of sprays simply as "aromatic hydrocarbons."

Drain and toilet cleaners

Air fresheners, often used in bathrooms to mask odors, do so in some cases by emitting compounds that diminish our ability to smell. Toxic chemicals such as cresol, which attacks the central nervous system, kidneys, liver, spleen, and pancreas, are commonly found in disinfectants. Aluminum chlorhydrate (or related aluminum compounds) are often the active ingredient in deodorants. Since its implication as a possible factor in Alzheimer's disease, aluminum is one element best avoided whenever possible. Drain and toilet cleaners, which are almost entirely sodium hydroxide (lye), a powerful toxin and mucous membrane irritant, are generally considered safe when used as directed. However, if a chlorine-based cleanser or bleach is accidentally used in conjunction with a drain opener containing ammonia, a deadly gas—chloramine—is produced.

First Steps

- **Keep toilets, tubs, and sinks clean and mold-free.**

- **Dry off tile walls around tub and shower to discourage microbial growth.**

- **Clean mold or mildew with a strong solution of vinegar and water.**

- **Substitute oxygen bleach cleansers for chlorine bleach cleansers.**

- **Avoid aerosol cans whenever possible.**

- **Discard any half-used products you no longer need.**

- **Avoid hair spray or other products with methylene chloride.**

- **Ventilate the bathroom especially well.**

- **Avoid the use of chemical air fresheners.**

- **Use chemical drain openers sparingly and carefully and never use in conjunction with any other cleaning product.**

Housecleaning

Mites and particulates

It may be hard to believe, but housecleaning—dusting, vacuuming, bed making—is a highly polluting activity. Sweeping and vacuuming, for example, may reduce the dust level on the floor or rugs but resuspends much of that dust in the air where the small particulates are likely to be inhaled. In the case of vacuuming, especially, the air sucked into the vacuum is obviously laden with dust, but as that air is exhausted through the paper or fabric bag which holds most of the dust, the smaller (and more deeply inhalable) particles are allowed to escape into the air, where their extreme lightness allows them to remain suspended for long periods. The composition of house dust varies from urban to suburban and country sites and from season to season, but it generally contains a mix of biological, organic, and inorganic agents, including pesticides, detergents, pollen, mites, viruses, bacteria, asbestos fibers, and human sebum—a prime vehicle for bacteria.

Allergies

An all-but-invisible human parasite—the house dust mite—is a potent source of allergic reaction. This mite lives off human skin scales, and its pulverized debris and excrement causes allergic reactions in sensitive individuals. Unlike lice or "crabs"—parasites connected with inadequate hygiene—this mite is a member of the fauna that almost universally inhabits the human body. House dust mites are most abundant in mattresses, bed linens, and upholstered furniture, and favor temperatures around 77 degrees Fahrenheit. They require a relative humidity above 45 percent. Their needs, therefore, fit in perfectly with those of their hosts.

First Steps

- **Replace regular portable vacuum cleaner with a central vacuum cleaning system, which carries all the dust into a main suction unit in the basement.**

- **Rather than vigorously shake out sheets and blankets daily (which monumentally increases particulate content of air and redistributes mites and dust), change bed linens frequently, even as**

often as twice or three times a week where there are allergic household members.

- **Vacuum upholstered furniture often. Don't just beat the dust out of it.**

- **Avoid spray products for dusting, polishing furniture, or spot-cleaning carpets. (See Product Substitution chart in chapter 4.)**

- **Use silicone-treated dustcloths, which hold the dust rather than resuspend it in the air.**

- **Substitute low-polluting cleaning products for high-polluting ones. (See Product Substitution chart.)**

- **Brush and groom hairy pets often—outdoors.**

The Basement

"Out of sight, out of mind" best describes most people's attitude about the basement. As long as the heat comes up, the water stays hot and plentiful, and no strange noises erupt from the depths, the basement is a rarely considered part of the home. But within that space is the potential for a lot of pollution that easily rises, unnoticed, into the rest of the home.

The furnace

Take, for example, the furnace. According to a 1987 survey done by the Housing and Urban Development department of the U.S. government, there are almost 103 million homes in the country (including both single- and multiple-family dwellings). Of these, about 54.4 million rely primarily on gas for heat and about 14 million use oil-fired furnaces. Another 24.4 million use electric heat, about 6.4 million still rely on wood, 471,000 use coal, and 1.2 million use kerosene or other fuels. More than half of the total number of homes have forced air heating systems which circulate air, warmed by the furnace, throughout the house via air ducts. Older homes may have steam or hot-water radiators.

With the exception of the all-electric homes, then, there is a constant production of combustion by-products within the confines of the furnace. Although virtually all gas and oil burners are vented outdoors, poorly maintained burners, faulty flues, cracks, and leaks in the pipes could cause a great deal of carbon monoxide, nitrogen dioxide, nitric oxide, aldehydes, and other organic compounds and fine particles to enter the home. The most imminent danger is that of carbon monoxide poisoning, which results in the death of almost 1500 people each year. Oil and gas companies, sensitive to this problem, will usually examine your furnace system on request, or give detailed instructions for you to check it yourself.

Sources of asbestos

A major source of asbestos also resides in the basement, in the form of insulation around water and heat pipes. The protective covering on this insulation may tend to tear or wear away with age, exposing the asbestos-containing material beneath it and freeing asbestos fibers to float through your environment. Ironically, the smaller and more respirable of these fibers have the greatest capacity for remaining airborne—some, up to twenty hours in still air and even longer in turbulent conditions, such as during use of fans or air conditioning systems.

Handling asbestos

Asbestos fibers are not only irritating to your respiratory system, skin, and eyes but are highly carcinogenic when those fibers penetrate to the pulmonary system. The fibers are sharp and highly durable, maintaining their physical integrity even when embedded in human tissue. For this reason, you should never attempt to remove or repair asbestos-containing material; you may inhale even more of the fibers or free them into the home. This is one job for experts.

First Steps

■ **If you have a gas- or oil-burning heating or hot water system, check burners for proper adjustment and check flues and other pipes for cracks and leaks. Repair as needed.**

■ **Replace combustion fuel systems with electrical systems.**

■ **Check air ducts for corrosion or any loose, fibrous material and repair where needed.**

■ **Install a ventilating heat exchanger (see chapter 4).**

■ Check for shredding or deteriorating pipe insulation and repair or replace. (If you suspect you have asbestos pipe insulation, get a professional to do the work.)

■ Seal other fibrous pipe insulation with a penetrant containing vinyl acrylic polymers and inorganic materials.

The Garage

Any attached garage used to store cars or other fueled vehicles such as lawnmowers, snow blowers, or tractors can be a source of carbon monoxide and other exhaust fumes if the vehicles are run within its confines. Idling cars in the garage on chilly mornings have resulted in many accidental deaths in bedrooms above or attached to the garage. Instead, cars started within the garage should be immediately driven outside to warm up. In constructing a new home, keep garage and house separate.

Building Superstructure

Wood, brick, stone, cement slab—the generous resources of this country make a wide range of building materials available to the residential housing market. But again, nature's bounty and industrial ingenuity are not without their price. Many of today's building materials bring with them the potential for an additional indoor pollution burden.

Plywood, particleboard, and fiberboard

As we have already discussed, plywood, particleboard, and fiberboard (but not plasterboard, an entirely different material) exude large quantities of formaldehyde, especially during the first months, or even years, of their use. In efficiently weathertight housing incorporating a great deal of these materials, this outgassing may increase the interior level of formaldehyde above a safe and healthful limit. The problem most seriously affects those who live in mobile homes, which are constructed mainly of such materials and must be heavily insulated and weather-proofed.

There are more than four million mobile homes in the United States, and whereas conventional homes have average formaldehyde levels below 0.07 ppm, mobile homes, particularly new ones, may have readings between .3 and 1.0 ppm. In 1981, the Environmental Protection Agency estimated that about 2,400,000 people in mobile homes are exposed to an average of up to 400 parts per billion (ppb), or .4 ppm, and another two million to average levels of up to 350 ppb, hundreds of times higher than ambient levels in nonurban areas.

Urea-formaldehyde foam insulation

Another source of formaldehyde that has received considerable attention recently is urea-formaldehyde foam insulation (UFFI), made by mixing UF resin with a foaming agent and an acid catalyst under pressure. The foam can be easily blown into the walls of buildings where it hardens within minutes and cures and dries within days. This ease of application made it ideal for homeowners of the late 1970s and early 1980s who found themselves under siege from burgeoning heating bills.

Approximately 170,000 houses were insulated with UFFI in 1977, its peak production year, and about 150,000 additional houses were retrofitted with it annually after that until it was discovered that in an overwhelming number of cases the formaldehyde was leaching from the product into the air inside the houses, often causing acute illness among residents.

The Consumer Product Safety Commission received thousands of complaints from residents of homes fitted with the insulation, claiming acute and chronic respiratory problems, headaches, and other neurological problems, and on February 22, 1982, it banned use of the foam as being a health threat. That ban was subsequently overturned by the U.S. Court of Appeals in April 1983, ruling that there was inadequate evidence.

Homeowners found themselves caught between the Internal Revenue Service's financial encouragement of home insulation in the form of a $300 tax credit, and fear that UFFI would adversely affect their family's health. The controversy remains, although residential use of the foam has dropped considerably.

Asbestos tile and cement

Another building material that could have a profound impact on the structure's residents is asbestos tile, widely used after World War II as a façade material in rural housing because of its low cost and durability. Examined at close range, these tiles can often be recognized by the wavy yet parallel scoring lines used over most of their surface to impart texture. They are generally quite hard and brittle, but if broken do look fibrous around the edges. They can be painted any color. As the material weathers, it may begin to shed asbestos fibers, some of which could enter the home through ventilation points, doors, and windows. Even more important, some homeowners do not identify the shingles as an asbestos product and in trying to remove or repair the structure, they inhale the fibers. Again, this calls for expert assistance.

Some apartment buildings, too, suffer from a hidden asbestos problem in the form of troweled-on acoustic or decorative material, most often in hallways and entranceways. If this material is disturbed by vigorous cleaning or vibrations, such as might be encountered during building renovation, it may shed asbestos fibers into the surrounding air. Unfortunately, it is virtually impossible for the average resident to accurately identify this sort of asbestos-containing material, as it is identical to non-asbestos-containing stuccos and cements.

Another ominous threat from asbestos has surfaced, as loose asbestos fibers have been found in drinking water that was transported by asbestos-cement-lined pipes. In one recent situation, the entire town of Woodstock, New York was warned against drinking its town-supplied water that was discovered to have been heavily contaminated by asbestos eroding from the pipeline.

Asbestos fibers contained in drinking water can lodge in the stomach or intestines causing irritation, tumors, and cancer.

Glass fiber

Asbestos may also be present in some insulation materials, such as those available in rolls of batting. More prevalent in these, however, is glass fiber, a fiber of extruded glass, hence the name. Although there is not sufficient evidence to link glass fiber with cancer or any other specific health problem, there is concern over potential health effects, based on the product's widespread and growing use and on some research into fiber carcinogenicity in animals.

Radon

Still another, and even more serious, threat exists in the form of radon 222, an inert but radioactive gas that naturally occurs in many kinds of rocks and soil throughout the country and that has been found in building materials such as concrete, tile, and brick. Radon has the ability to migrate from the material it is in, giving it plenty of opportunity to enter your indoor environment. It is the short-lived decay products of radon—polonium 218, lead 214, bismuth 214, and polonium 214—which emit significant doses of alpha radiation.

Although we are constantly bombarded by very low-level natural radiation of various kinds, researchers blame indoor radon levels for 10,000 to 20,000 additional lung cancer deaths per year, making indoor exposure to radon progeny responsible for most of the cancer deaths among nonsmokers. Officials of the EPA have publicly

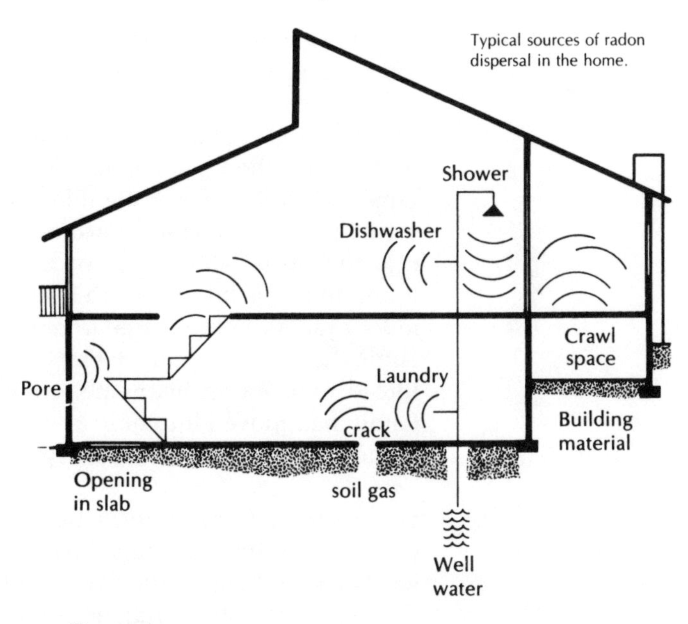

Figure A.

stressed that radon exposure is the most serious radiation threat we face today and that as a national problem it is more serious than the recent Love Canal toxic waste crisis.

How radon enters

There are basically three ways radon gas enters the home. The most significant is by seepage from the soil around and under the building through cracks and openings in the structure itself. As noted, it may also emanate from earth-derived building products, especially those that may be composed of waste products of industrial processes such as bricks or concrete containing phosphate slag, a by-product of phosphate production. The third way radon can enter your home is via well water or even natural gas used for cooking and heating.

In early 1985, Stanley J. Watras, an engineer working for the Philadelphia Electric Company's Limerick nuclear power plant in eastern Pennsylvania, set off a radiation alarm when he stepped into the plant. The source of his radiation was found to be his home, which boasted the highest indoor radiation levels due to natural radon contamination ever found in the U.S. Watras was forced to abandon his home, which he and his family had moved into only a year earlier. Other homes in the area were similarly contaminated.[16]

Some areas of the country have a higher incidence of radon problems than others. Certain locations in Maine, Illinois, Montana, eastern Pennsylvania, Maryland, northern New Jersey, and New York state have pockets of rockbed high in radon concentration. Some scientists, such as Dr. Bernard Cohen, a physics professor at the University of Pittsburgh and a leading authority on radon, believe that virtually every state has areas of natural radon contamination that might pose a threat to residents. Other sites of concern include residential properties built over now-deserted mining areas such as those in Florida and Tennessee and on uranium tailings such as those in Colorado.

Radon progeny

Radon gas emanating in the open air does not pose a major health threat, because it quickly disperses. In the home, however, it collects, and the radiation-emitting radon progeny linger in the air breathed by the residents. The electrically charged progeny attach themselves to the respirable particles floating freely in the air and are subsequently inhaled. These progeny can lodge themselves deeply into the lungs, where their alpha radiation is released. Over time this radiation can cause cancer. Efforts to make homes more airtight exacerbate the problem by increasing the radon accumulation. In the simplest example, it can be considered that reducing the ventilation rate by half would just about double the radon level.

The serious and pervasive nature of radon gas as an indoor pollutant has encouraged the U.S. Environmental Protection Agency and some states to offer radon testing services. In addition, radon detectors for use in the home are available from commercial sources. These radon detectors are generally small devices that remain in the home for about a week and are then sent back to the company for analysis. (See Appendix for specific sources for radon detection services.)

Vapor barrier

Traditionally, many wood-frame houses were wrapped with tar paper or a similar material to increase their weather-and water-resistance. Recent technology has brought an improvement on this in the form of synthetic building wraps, such as Du Pont's Tyvek, which provide even greater protection and energy savings. This material is designed to allow for only slight air permeability, which has the effect of drastically decreasing infiltration and exfiltration and diffusion of contaminants out of the residence. In buildings with inadequate ventilation, such a building wrap may also lead to dampness from a buildup of water vapor unable to escape. This can contribute to mold and mildew growth in areas of the house and in wall and ceiling insulation.

First Steps

■ Seal all cracks in basement walls and concrete slabs with polymeric caulks to prevent infiltration of radon gas.

■ Ventilate crawl spaces to dispel radon gas.

■ In areas known to have a radon problem:

Cover basement walls and concrete slabs with epoxy paint, polymeric sealant, or polyurethane or polyamide film vapor barrier to prevent radon gas from entering the building through pores in these materials.

Before constructing a house, excavate local soil beneath foundation, and replace it with low-radium-content soil.

Use an effective air filtration system indoors to reduce radon progeny (see chapter 4 for details).

What to do

■ Do not build in areas of uranium or phosphate mining or where tailings have been used for landfill.

■ **If you believe you live in an area prone to radon gas, seek professional testing by government or commercial sources.**

■ **Substitute solid wood for plywood building elements (such as subfloors) whenever possible.**

■ **Where plywood or particleboard is used, cover it with shellac, varnish, polyurethane, or other diffusion barrier.**

■ **In construction, be certain there is sufficient ventilation when making use of vapor barrier building wraps.**

■ **Avoid the use of urea-formaldehyde foam insulation.**

■ **If you believe your house's façade is asbestos tile, *do not attempt to remove it yourself*. Consult a professional to determine if it should be removed (if it's in good shape, removal may not be necessary). Use a professional, too, for exterior painting.**

Miscellaneous

Some sources of indoor pollution are not limited to one room or area of the home. They are either mobile with the residents or possible to locate almost anywhere. One source of pollution that falls into this category can also be considered the single most prevalent and dangerous health threat today—that is, tobacco smoke.

Tobacco smoke

According to the 1989 U.S. Surgeon General's Report, an estimated 574 billion cigarettes were consumed in 1987, or 3,196 cigarettes per smoker. Although this number may seem shocking, it is actually the lowest recorded per capita consumption since the 1940s and the lowest total consumption since 1972. And although the percentage of American adults (aged 20 years and older) that smoke has been slowly declining since 1966, the fact that more than 29 percent of our population still engages in this habit means that it is unlikely that anyone can avoid exposure to cigarette smoke. "Passive exposure" will be encountered at sporting events and in public transportation, offices, restaurants, homes, industrial facilities, and at almost any other location considered.

In one major study on the health effects of air pollution, it was discovered that an average of 70 percent of homes in eight target cities housed at least one smoker. Smoking in the home has been found to impair the lung functioning of children and to increase the number of "restricted-activity days" and "bed-disability days" in rough proportion to the number of smokers per household or the number of cigarettes smoked.[17]

Mainstream and sidestream smoke

With regard to the workplace, William L. Weiss of Seattle University School of Business concluded that a business loses $5,600 per year per smoker and that $560 of that loss is due to adverse health effects felt by nonsmokers exposed to cigarette smoke on the job. Nonsmokers who work on a job with smokers are subject to the same health effects as light smokers—most notably, impaired pulmonary functions and increased respiratory illness. In addition, nonsmokers forced to work in a smoke-laden environment have additional reasons for job dissatisfaction that could lead to lowered productivity or to quitting.[18]

Clearly, there is reason for concern. Tobacco smoke comprises more than 2,000 chemical compounds, at least forty of which are proven to be cancer causing (see Table F). It may be argued that there are actually two distinct types of cigarette smoke: mainstream smoke, which is undiluted and pulled directly through the tobacco into the smoker's lungs; and sidestream smoke, the smoke as it exits from the cigarette into the surrounding air. The proportional composition of the two are quite different, with sidestream smoke actually much higher in the critical compounds.

Passive smoking

Still, the passive smoker, breathing diluted sidestream smoke, does not get the lung dose that the smoker absorbs. A typical cigarette smoker inhales mainstream smoke 8 to 10 times per cigarette and absorbs from 50 percent to 90 percent of the water soluble compounds and 20 percent to 70 percent of the nonsoluble compounds present. But sidestream smoke is produced during 96 percent of the total burn time of each cigarette, greatly increasing the nonsmoker's exposure to

Table F: Partial Composition of Cigarette Smoke

Particles:	Gases and Vapors:
Tar (chloroform extract)	Water
Nicotine[T]	Carbon monoxide[T]
Benzo(a)pyrene[C]	Ammonia[T]
Pyrene[*]	Carbon dioxide
Fluoranthene[*]	Nitrogen oxides
Benzo(a)fluorene	Hydrogen cyanide[T]
Benzo(b/c)fluorene	Hydrogen sulfide
Chrysene, benz(a)anthracene[C]	Acrolein
Benzo(b/k/j)fluoranthene[C]	Formaldehyde[I,T]
Benzo(e)pyrene[C]	Acetaldehyde[I,T]
Perylene	Toluene[I]
Dibenz(a,j)anthracene[C]	Acetone[I]
Dibenz(a,h)anthracene, ideno-(2,3-ed)pyrene[C]	Polonium 210[R]
Benzo(ghi)perylene	Methane, ethane, propane, butane, etc.[H*]
Anthanthrene	Acetylene, ethylene, propylene, etc.[H*]
Phenols[*]	Methanol
Cadmium[T]	Methyl ethyl ketone[I]
Nitrosamines[C]	Methyl nitrate
	Methyl chloride

[C]Carcinogen; [*]suspected carcinogen or co-carcinogen
[T]Toxin
[I]Mucous membrane irritant
[H]Hydrocarbons, many of which are toxic or under investigation as suspected carcinogens, mutagens, teratogens, etc. All are fat soluble and accumulate in human adipose tissue where they can constitute a substantial body burden.

Adapted from National Research Council, *Indoor Pollutants,* p. IV 101.

respirable particles, carbon monoxide, benzo(a)pyrene, nicotine, acrolein, and other undesirable chemicals.

Where the ventilation rate is low, the increases can be dramatic. It is estimated that six cigarette smokers at a party can easily produce particulate levels of 900 micrograms per cubic meter, or the air emergency level.[19] The particulate levels are especially important because respirable particulates such as those produced in tobacco smoke are often the vehicles by which radon progeny transport themselves deep into the lungs.

Studies both in the U.S. and in Japan indicate that the incidence of lung cancer in nonsmoking women is two to four times higher for those who were married to smoking men. The rate of emphysema was 1.4 times higher for that same population in the Japanese study.[20]

First Steps

- **Prohibit tobacco smoking in your home or limit it to one well-ventilated room.**

- **Use an effective air cleaner in all smoking areas. (See chapter 4 for more information.)**

Hobbies

Hobbies and do-it-yourself projects also contribute to the home pollution load, mainly in the form of airborne particulates and chemical fumes (see Table G). Woodworking, one of the most popular home projects, often requires large amounts of sanding, which releases sawdust. Although most of this sawdust is heavy enough to fall from the air straight away, a certain amount of lightweight dust remains airborne and can be inhaled into the respiratory system, where it may be a major irritant. Woodstains, paints, paint strippers, turpentine, shellac, varnish, and polyurethane all contain an array of organic chemicals and other compounds (such as hydro-

Table G: Active Ingredients and Selected Health Effects of Common Hobby Materials

Hobby	Product/Activity	Pollutant/Compound	Selected Health Effects*
Woodworking	Wood sanding	Sawdust (airborne particulates)	Eye and nose irritation
	Protective coating	Polyurethane	
	Woodstain	Hydrocarbons, ether	Narcotic effects
	Varnish	Petroleum distillates	
	Paint remover	Methylene chloride	Carcinogenic; causes increased carboxyhemoglobin levels; may cause narcosis and brain cell damage at high levels
	Paint thinner	Naphthalene, turpentine	Respiratory irritation, nausea
	Lacquer (solvents)	Benzene, toluene, xylol, xylene, ethyl ether, acetone	Skin absorption can produce chronic poisoning; nasal irritation, nausea, narcosis, and neurological effects at moderate levels; carcinogenic; may cause birth defects
	Shellac (thinner)	Methanol	Repeated exposure may result in cumulative body concentration
Metalworking	Plating	Hydrocarbons, resins	Flammable; highly irritating to eyes, skin, and respiratory system
Graphics	Adhesives	Petroleum distillates aerosol spray propellents	
	Marking pens	Hydrocarbons, petroleum distillates	
Jewelry, stained glass	Soldering	Lead	Toxic; may cause brain damage
	Glass frosting	Toluene, xylene	Skin and mucous membrane irritation; can cause birth defects or fetal death; central nervous system depressant

* For a more complete discussion of adverse health effects, see Organic Chemicals section of the Pollution Compendium, chapter 6.

carbons, resins, toluol, methylene chloride, naphthalene, benzene, acetone, ethyl ether, xylene), which are liberated into the indoor environment both during use and during storage of previously opened containers. In addition, there is generally a long, slow drying and curing process of these materials on the finished project, during which chemicals are outgassed into the room.

Vapors and aerosols

Model building (railroads, airplanes, ships, birds) also require the use of glue and paints that release compounds such as petroleum distillates, which are best left out of the home. Many amateur graphic artists use marking pens and rubber cement, which also release these solvent fumes. Spray adhesive, a convenient medium for attaching paper, cardboard, plastic and metal films, and photographs, is especially dangerous in that it expels a cloud of fine aerosol of these compounds directly into the air, along with a sizeable dose of propellents. This aerosol is highly flammable along with being extremely toxic to eyes, skin, and mucous membranes. The warning labels on one such spray adhesive can reads: "DANGER: Extremely flammable. Keep from heat or flame. Contents under pressure. Vapor is harmful. Use with adequate ventilation. Avoid breathing of vapor or spray mist and prolonged contact with skin. KEEP AWAY FROM CHILDREN."

Some less common hobbies, such as jewelry making or stained glass working, are also sources of indoor pollution. The flux and solder used during soldering of leaded glass work and some jewelry releases fumes laden with lead (which is toxic), hydrocarbons, and resins, all of which may cause such acute effects as dizziness, breathing difficulties, and mucous membrane irritation.

Proper ventilation

Of course, it would be ridiculous to suggest that people eliminate crafts and hobbies from their lives. Instead, it is important to isolate these activities from the mainstream of the home or apartment and to provide adequate ventilation in the areas where they are performed. This can be accomplished by as simple a means as opening a nearby window and providing an exhaust fan, or as technical and efficient a way as installing a special outside-vented air duct (similar to the range hood over a gas stove). Either way, the effect will be to allow the hobbyist to enjoy his or her work and living space in safety.

First Steps

- **Do all wood sanding and use of solvents and fluxes in a well-ventilated room separate from the rest of the house.**

- **Seal partially used paints, solvents, strippers, and other chemical compounds tightly and store in a well-ventilated area away from air ducts and heat.**

■ **Discard any such products that will not be used again.**

A note to campers

After a week at the office or behind a school desk, a walk through the woods in clear sunshine and fresh, unspoiled air can seem most inviting and refreshing to body and spirit. No doubt it is, unless you end the day by cooking in your tent beneath the warm glow of the kerosene lantern. These unvented combustion appliances of the open road threaten their users with significant NO_2, CO, CO_2, and particulate exposure when used in an enclosed space (such as a tent or small cabin). Emission factors for two commonly used camping appliances—an Optimus 123 stove and a model 288–700 Coleman lantern—have been determined in a laboratory study. Both appliances were run on white gas (Coleman fuel) at their highest heat or light output. The stove was tested with an aluminum pot of water on the burner. The results can be seen in Table H.

Air exchange rates were calculated for small two-person tents and were determined to be 85 to 130 exchanges per hour. At this rate, concentrations of 25 ppm CO, 580 ppb NO_2, and 360 ppb NO were estimated if a stove of the type tested was used in the tent. Actual concentrations of CO inside tents with stoves running have been measured by one researcher at 30 ppm, and by another between 10 ppm and 100 ppm. As you can see, no place is entirely safe from the threat of indoor pollution. The solution remains the same, however—switch to electric flashlights for indoor use, cook outside the tent, or in an open alcove, and ventilate the tent as much as is practical.[21]

Table H: Gas Stove and Lantern Emissions (Mean values +/− standard deviation)

	Fuel Consumption (kg/hr)	CO (g/kg)	NO_2(g/kg)	NO (g/kg)	NO/NO_2
Stove	0.124	2.71 +/− 4.28	1.03 +/− 0.04	0.35 +/− 0.9	0.34 +/− 0.02
Lantern	0.056	*	0.71 +/− 0.09	5.33 +/− 0.5	7.58 +/− 1.00

* Preliminary measurements of CO for the lantern found levels close to background levels; therefore CO measurements for the lantern were not taken.

Note: When the stove was turned down to low, CO and NO_2 output were halved and NO went down to barely above background level. When the lantern was turned down to low, both NO_2 and NO went down to just above background level.

Source: Spengler, John D., and Cohen, Martin A., "Emissions from Indoor Combustion Sources," *Indoor Air and Human Health*, Gammage, R.B., and Kaye, S.V., eds., pp. 271-273.

Pesticides

The scenario is familiar to many city dwellers:

You wake in the middle of the night, hungry or thirsty, and head for the refrigerator for a drink and a snack to satisfy the urge. You flip on the light, and the kitchen counter seems to come alive with a hoard of small, scurrying creatures. Your peace of mind is ruined: You've got roaches.

Those who live in the country are plagued less by these than by flies, ants, termites, rodents, and so on. But the problem of indoor pests is virtually universal. Unfortunately, so is our dependence on chemical pesticides, which are used in 90 percent of all North American homes. These can persist in our home environment well beyond the time needed to do their job—in fact this persistence is an intrinsic quality of most pesticides. An excellent example is the popular No-Pest insecticide strips, made to be hung indoors, that saturate the air with powerful chemicals for several weeks at a time. Among the ingredients of these strips is dichlorvos (2–2 dichlorovinyl dimethyl phosphate), which is under investigation by the EPA for possible oncogenic, mutagenic, teratogenic, fetotoxic, and neurotoxic effects.

Chlordane

The most widely used insecticide against termites, chlordane, persists for up to five years in soil or wood products. In 1975 almost 25 million pounds were used for pest control, but by 1979 its manufacture here was suspended because of its overwhelming toxicity and its ability to penetrate into the home and bodies of residents. In some cases, people were forced to abandon their homes when high levels of the chemical were discovered indoors. Forced-air heating systems contribute greatly to pesticide dispersal from the basement and crawl space up into the home. In some cases, pesticides applied indoors during the warm weather months have been unwittingly resuspended in the air when the heating system was started up in later months. (See Table I for active ingredients of some common pesticides.)

First Steps

■ **Eliminate need for future termite extermination by using termite-proof foundations, such as**

Table I: Active Ingredients in Some Common Pesticides

Note that virtually all of the listed ingredients can be classified as toxic or producing adverse health effects if ingested or inhaled in high concentrations or over prolonged periods. In general, however, they are currently regarded as safe for their intended use in a well-ventilated area. Decreased ventilation could well have health implications.

Target pests	Active Ingredients
Wasps and hornets	O-isopropoxphenylmethyl carbamate, dichlorvos, petroleum distillates
Ants and roaches	Pyrethrins, piperonyl butoxide, n-octyl sulfoxide of isosafrole, petroleum distillates, kepone, mirex diazinon, propoxur
Mosquitoes and flies	Dicarboxamide, petroleum distillates, propoxur
Moths	Dicarboxamide, petroleum distillates
General insects	Cyclopropane, petroleum distillates
General insects	Toluamide, dichlorvos (Vapona in pest strips)
Cat fleas	Carboxytheyl carbamate
Dog fleas	Dichlorvos
Termites, fire ants	Chlordane (discontinued in 1979), heptachlor, aldrin, dieldrin, lindane
Plant pests	Malathion, Kelthane, xylene, carbaryl (Sevin), Meta-systox R

Adapted from Meyer, Beat, *Indoor Air Quality*, Addison-Wesley, Reading, MA, 1983.

concrete block, poured concrete, or fieldstone in new constructions.

- **Replace indoor insecticides with bug traps or non-polluting pesticides. (See Product Substitution Table in chapter 4.)**

- **To rid houseplants of pests, wash plants often in warm soapy water. Rinse with clear water.**

- **Especially avoid aerosol spray insecticides.**

Lighting fixtures

Two pollutants—ozone and polychlorinated biphenyls (PCBs)—may be emitted by certain types of lighting fixtures in the home. Although mercury-enhanced indoor lightbulbs are not yet in common use, the U.S. Department of Energy has proposed a plan to promote them as being more energy efficient. These bulbs, which can be recognized on city streets by their pinkish light, do emit ozone, which would significantly affect its concentration indoors. Although ozone provides a critical service in the upper layers of our atmosphere by filtering ultraviolet light, which might otherwise reach life-threatening levels, it is highly irritating to mucous membranes and can even be toxic at high concentrations.

Many pre-1978 fluorescent fixtures have capacitors in their ballasts (the starting mechanisms) that contain PCBs, highly toxic substances that have been observed indoors at levels up to thousands of times higher than ambient (outdoor) levels. These outrageously high concentrations were always found in kitchens and bathrooms with PCB-filled fluorescent light ballasts. Removal of these ballasts caused the levels to drop dramatically within two months.

First Steps

- **Replace old fluorescent fixtures with new, preferably incandescent, lighting.**

Controlling pollution

Although this house tour has pointed out a plethora of pollutants capable of producing symptoms from itchy eyes to cancer, its purpose is not to alarm or depress you, but to alert you to a problem you can help control. The accompanying "Walk Through the House Checklist" should help you develop your home's personal pollution profile. Once you know the trouble areas—whether it's the kitchen with its gas appliances, an over-abundance of spray products, the pervasive pall of tobacco smoke, or even the ominous threat of radon gas—you can begin taking steps to substantially lower pollutant levels. You and your family can live healthier.

The Walk Through the House Checklist

For your home's personal pollution profile, check off all applicable pollution sources for your score. Then add the numbers in parentheses after each item.

The Kitchen

□ Unvented gas stove (3)

□ Unvented gas oven (3)

□ Unvented gas clothes dryer (3)

□ Unvented electric clothes dryer (2)

Aerosols (used regularly):
□ Oven cleaner (1)
□ Air freshener (1)
□ Water and stain repellent (1)
□ Disinfectant (1)
□ Others (add 1/2 point for each additional product)

□ Scouring powder (with chlorine bleach) (1/2)

□ Ammonia (1)

□ General purpose cleaners (e.g. Fantastik) (1/2)

□ Spot remover (1)

□ Floor wax (self-cleaning, e.g. Wood Preen) (1)

The Living Room

- ☐ Fireplace or wood stove (4)
- ☐ Coal stove (4)
- ☐ Wall-to-wall carpeting (and pads) (2)
- ☐ Carpet shampoo (1)
- ☐ Decorative paneling or partition walls (new: 3, 1+ years: 2, 3+ years: 1)
- ☐ Furniture or shelving made with plywood or particleboard (new: 2, 2+ years: 1)
- ☐ Upholstered furniture (1)

Bedrooms and Nursery

- ☐ Unvented kerosene or gas space heaters (4)
- ☐ Cool-mist humidifier (2)
- ☐ Air conditioner (1 each)

The Bathroom

- ☐ Enclosed stall shower (1)

Aerosols:
- ☐ Hair spray (1/2)
- ☐ Deodorant (1/2)
- ☐ Air freshener (1)
- ☐ Shaving cream (1/2)
- ☐ Tub and tile cleaner (1)
- ☐ Disinfectant (1)

- ☐ Chlorine bleach-based scouring powder (1/2)

Housekeeping	□ Do you clean house (dust, sweep, vacuum, shake out sheets and blankets) regularly—at least twice a week? (1)
	□ Portable vacuum cleaner (1)
	□ Regular use of dusting spray (1)
	□ Regular use of furniture polish (1)
The Basement	□ Gas or oil furnace (2)
	□ Gas or oil hot water heater (2)
	□ Asbestos pipe insulation (3)
	□ Unfiltered forced-air heating system (2)
The Attached Garage	□ Motor vehicles (4)
Building Superstructure	□ Plywood, particleboard, fiberboard (in new homes) (3)
	□ Urea-formaldehyde foam insulation (new: 4, 1+ years: 3, 5+ years: 2)
	□ Glass fiber insulation in attic or below ground floor (2)
	□ Vapor barrier (e.g. Tyvek) (1)
	□ Asbestos shingle façade (1)

- ☐ Double- or triple-glazed (thermal) windows (2)

Foundation/ Subsoil

- ☐ Do you live in an area known to have (or suspected of having) high radon content in the soil or bedrock? (4)
- ☐ Have you had your foundation and surrounding soil treated for termites, ants, or flies within the past five years? (3)

Water Supply

In a radon-prone area:

- ☐ Is your water supplied by a private well? (3)

Miscellaneous

- ☐ Do you or any members of your family smoke? (4 per smoker)
- ☐ Are you often visited by smokers? (3)
- ☐ Do any furry pets or birds live with you? (1)
- ☐ Do you have any house plants? (1)
- ☐ Are insecticides or pesticides used indoors? (regularly: 2, rarely: 1)

Hobbies:

- ☐ Woodworking, carpentry (2)
- ☐ Model building (1)
- ☐ Jewelry or lamp-making (soldering) (2)
- ☐ Graphic arts (use of spray adhesives, marking pens, etc.) (1)

- ☐ Are there pre-1978 fluorescent lighting fixtures anywhere in the house? (1)

Symptoms

- ☐ Do you store old, partially used cans of paint, solvents, mothballs, pesticides, polyurethane, paint stripper, etc., in the house? (1)
- ☐ Does you house often feel damp, especially in winter? (2)
- ☐ Are there any unusual or foul odors in places in your house? (1)
- ☐ Do cooking and other odors tend to linger? (1)

Your score to this point ____________

Methods of Improving Air Quality

(Subtract the number of points indicated after each item from the above score.)

- ☐ Do you use an ion generator, electrostatic precipitator, or effective air filter for indoor air cleaning? (4 points per unit)
- ☐ Do you use and maintain an effective air filter in a forced-air heating system? (4)
- ☐ Do you open windows for ventilation in all seasons? (4)
- ☐ Do you use a heat exchanger to provide ventilation without sacrificing energy-efficiency? (6)

- ☐ Do you use a central vacuum cleaning system? (2)

Your total indoor air quality score ———

Understanding Your Score

0–15 Exceptionally high indoor air quality.

16–25 Reasonably safe indoor air, with room for improvement, especially if there are unvented fueled appliances or a threat of radon.

26–50 Family members who spend long periods indoors or who tend to be allergic or have other health sensitivities may suffer adverse effects. Acute health effects such as respiratory symptoms, irritability, or fatigue may be apparent. Chronic health problems and long-range illnesses may have their silent beginnings.

50+ Poor indoor air quality which could easily lead to both long- and short-range health problems. With a score this high, however, it should be relatively easy to take steps toward improvement.

3 A Walk Through the Office; A Stop at the School

■ In the summer of 1980, county health department officials were called in to investigate an outbreak of respiratory symptoms (dry throat, coughing, nasal congestion), eye irritation, and headache among office workers in a city office building. No apparent cause was discovered and the symptoms remained unexplained. It was only after the affected employees realized that the coughing and sensation of dry throat were exacerbated by holes being drilled through the carpet and floor (to install a cable for a new computer system) that the cause was discovered: Frequent carpet shampooing with a detergent concentrate was leaving a powdery residue that was responsible for the symptoms.[1]

A variety of symptoms

■ All the employees in three adjacent departments in a department store had experienced itching, a transient rash, and irritation of the eyes and respiratory system. Two other store employees had similar symptoms, but only when they worked short times in one of the affected departments. A fibrous dust left on glass counters was reported to officials who were investigating the outbreak. The cause was traced to the conduits of the air handling system that the three departments shared. Unlined matted glass fiber insulation had sustained repeated water damage from a malfunctioning water condensate tray in the unit. Fibrous glass particles released from the damaged ductwork at air levels well below the occupational standards were responsible for the symptoms, probably by means of physical contact with employees' hands and faces.[2]

■ Teachers and students at an elementary school were forced to vacate their classroom by noon each day because of widespread attacks of headache and nausea. An exhaust fan, inadvertently reversed during routine maintenance, was found to be functioning as an intake fan, bringing high levels of carbon monoxide into the classroom from the school's boiler stack.[3]

Legionnaire's disease

■ Health officials feared an outbreak of Legionnaire's disease when more than half of Itel's almost 200 workers began registering complaints ranging from headache, eye irritation, and fatigue to nausea, dizziness, and throat irritation, after the company moved them into a brand-new building in Port Washington, New York. When no bacterial contamination was found, the investigators tended toward a diagnosis of mass hysteria. In fact, total hydrocarbon concentration within the building was found to be about 1,600 to 1,650 micrograms/m^3 (as opposed to 210 +/ −60 micrograms/m^3 outdoors), and carbon dioxide levels had an eight-hour average of 1,000 ppm (peak levels were as high as 2,000 ppm) when the ventilation system was in an 85 percent recirculating mode (that is, only 15 percent fresh, outside air). Increasing the fresh-air intake alleviated virtually all the symptoms.[4]

Office buildings and schools have been among the first victims of the "sick-building syndrome," the elusive ailment that plagues buildings with energy-efficient heating, ventilation, and cooling (HVAC) systems, causing outbreaks of mysterious shared illnesses among those who spend time in the building. Complaints of building-related illnesses have been on the increase since 1978. Prior to that year, the National Institute for Occupational Safety and Health (NIOSH) had only performed six investigations of buildings for building related illness. Between 1978 and 1990 the agency conducted more than 600 such investigations. Between 1978 and 1980, NIOSH noted, 7.4 percent of its building evaluations related to indoor air quality and illness. In the 1980s that figure rose to 12 percent, and the current rate is about 20 percent of evaluations. These related mainly to government offices, business offices, schools, and colleges. In most of these investigations of mucous membrane irritation, headache, and fatigue, no specific causative agent was found, although increasing ventilation relieved most of the symptoms. Some 70 percent of the buildings involved were hermetically sealed and had largely recirculating HVAC systems.

Many causes

Naming a specific causative agent in cases of building-related epidemics is no easy task. In most cases, the suspected chemical and physical agents sampled (including virus or bacteria, molds or fungus, carbon monoxide, carbon dioxide, hydrocarbons, ozone, asbestos, glass fibers, formaldehyde, pesticides, and solvents) fall within occupational safety limits. When measurement of the airborne chemicals turns up no single contaminant in a concentration believed high enough to cause the symptoms, the investigators often consider the complaints to be without cause and the problem nonexistent or a product of mass hysteria. Much evidence exists, however, that numerous low-level exposures to a variety of common pollutants may have an additive or synergistic effect responsible for the onset of symptoms.

In many cases, the victims of the symptoms are the ones to point investigators in the direction of the cause. Sufferers are likely to associate certain rooms or areas in the building with an increase in their symptoms, or they remember the onset of symptoms in conjunction with some physical event that took place in the building (such as installation or cleaning of carpets, building maintenance, or construction). A tour of the prototype office and school will indicate typical generators of pollution indigenous to such buildings.

Building Materials

Outgassing

As you have seen in the walk through the house, many common building materials are sources of indoor pollutants that could achieve high concentrations when ventilation is inadequate. New building materials have been found to be a source of organic chemical contaminants because they contain residual solvents and other compounds from the manufacturing process. In a hypothetical office setting, researchers at Lawrence Berkeley Laboratory found that the peak pollutants emitted by new building materials, caulking, and carpeting included aliphatic hydrocarbons, toluene, alkylated benzenes, ketonic solvents, and specialty compounds such as butylated hydroxy-toluene (BHT). The outgassing of these solvents can be expected to diminish with time as the materials dry out. Formaldehyde is generally considered the only pollutant that is emitted by building materials over an extended period of time.[5] Office furniture, decorative paneling, carpets, and carpet pads are common sources of formaldehyde in the office setting.

One potential way of alleviating the problem of outgassing of hydrocarbons that are in new building materials is to allow these materials an effective "drying-out" period before they are used. Additional research is needed, however, to define acceptable drying-out periods for various materials. Another way of abating the problem might be to select low-emitting materials. Again, more research is needed, as is industry cooperation, in developing a practical range for such products.

Formaldehyde and asbestos

Older office buildings and schools may also be subject to invasion by fugitive asbestos fibers emanating from unlined insulation around pipes or in air ducts, or freed from ceiling panels, plaster, or fireproofing material by vigorous cleaning, maintenance, or physical damage. This has been considered a major problem in schools in New York, California, Maryland, New Jersey, Connecticut, and Massachusetts. In the San Francisco Federal Building, asbestos fibers from fireproofing material sprayed on structural beams between floors was discovered circulating through the 20-story air-conditioning system serving some 4,300 government employees. Although asbestos concentration in the offices measured below the OSHA standard of two fibers per cubic centimeter, a health hazard was declared for building workers. When asbestos fibers have been found floating in indoor air, removing the asbestos or isolating it in encasing materials has been the solution of choice. However, care must be taken in these procedures to avoid freeing additional fibers into the air.

Office Machines

Photocopying machines have long been recognized as potent sources of ozone, and wet-process copiers are also sources of organic chemicals and ammonia. The toner and developer in wet process machines are nearly pure aliphatic hydrocarbon petroleum distillate solvents that readily evaporate into the surrounding air and can cause acute respiratory discomfort. In addition, there have been some reports that some toners contain trace amounts of mutagens (birth-defect-causing substances) such as nitropyrene.

Ozone and copiers

Based on a survey of users, it is estimated that a wet-process photocopier is likely to use approximately 1 quart (1,000 grams) of combined fluids per week, giving a hypothetical emission rate of 25 grams per office per hour. Given a ventilation rate of one air change per hour (1 ach), the concentration by the end of the day will have nearly dissipated. However, at 0.2 ach (not uncommon in energy-efficient buildings), the concentration rises continuously during the day and peaks at 35 mg/m^3. At that same rate, there is barely time to remove the accumulated contaminants overnight.[6] Fortunately, wet-process photocopiers are becoming increasingly rare.

However, even the more advanced dry copiers emit ozone, and at a prodigious rate. A recently serviced machine produces about 4 micrograms per copy. After extended use the production can increase to up to 131 micrograms/copy, with an average of about 40 micrograms/copy. This results in breathing-level ozone concentrations of about 4–300 $micrograms/m^3$ for the operator.[7] Operator breathing-zone concentrations of up to 0.068 ppm have been measured under normal working conditions.[8]

Ozone's effects

The effects of ozone on humans range from simple respiratory and eye irritation, and less efficient functioning of lungs to asthma and coughing attacks and severe chest pains. According to the American Conference of Governmental Industrial Hygienists, the recommended occupational limit on ozone is 0.1 ppm or 200 mg/m^3 averaged over a normal eight-hour day or 40-hour work week. The current federal standard for outdoor air bars ozone concentration from exceeding 0.12 ppm more than once a year (although it is not uncommon for large metropolitan areas to fail to meet that standard). These limits were adopted at a time when it was believed that health effects began to occur at exposures of 0.24 ppm. Recent scientific studies indicate, however, that adverse health effects can begin at concentrations of 0.12 ppm and possibly lower (especially where there is prolonged or long-term exposure), and there is strong pressure on the government to revise its standard to as low as 0.08 ppm.

Reducing the ventilation rate and recycling more of the air inside office buildings exacerbates the buildup of ozone concentrations, especially in the copy rooms of companies with several machines running at a constant clip. A steady source of clean, fresh air is needed to dilute the ozone level to safe limits. Workers in hermetically sealed buildings are, of course, unable to introduce fresh air, since they cannot open the windows. But pressure on company and building managers may result in increased mechanical ventilation being provided.

Lighting

The effects of varying types of lighting on human beings have only recently received serious attention. Our ancestors rose with the sun, worked outdoors or in rooms with windows, and slept when it got dark. Today we are increasingly dependent on artificial lighting to keep our round-the-clock industrial economy in gear. Although more office buildings are constructed of hermetically sealed glass walls, fewer workers ever see the light of day during working hours. The windowed office is most often reserved for high-level executives, while the staff labors under the glow of fluorescent fixtures. (This, by the way, is no longer the case in Scandinavian countries, where workers, by law, are being guaranteed minimum amounts of natural light.)

Sunlight's benefits

But the spectrum of sunlight apparently has qualities we can ill afford to deprive ourselves of, according to researchers such as Dr. Richard Wurtman, a neuroendocrinologist at the Massachusetts Institute of Technology and a pioneer in light studies. He believes that the problem of light deprivation may afflict millions of working adults during the winter months, possibly triggering increased secretion of the hormone melatonin, which is normally released at night and turned off during the day. He relates this hormonal imbalance to seasonal bouts of depression, increased appetite, and craving for carbohydrates that attack some people during the winter. Another researcher, Dr. Alfred Lewy, of the University of Oregon Health Sciences Center, and his former colleague at the National Institute of Mental Health, Dr. Norman E. Rosenthal, reported significant success rates in treating victims of this disorder (known as seasonal affective disorder, or SAD) and similar disorders with exposure to ultra-bright lights for several hours each day.[9]

UV light

Ultraviolet light (UV), the shorter, nonvisible wavelengths of the spectrum that are responsible for the tanning and burning effects of sunlight, are also being used therapeutically. In an experimental treatment for some forms of leukemia and some autoimmune diseases, UV light is being used to trigger desired effects in the blood, alone or in conjunction with light-sensitive drugs.[10]

As for the visual effectiveness of artificial lighting versus full-spectrum sunlight, Dr. Richard H. Blackwell of Ohio State University believes that today's commonly used indoor lighting actually diminishes and impairs our effectiveness in performing a wide array of the visual tasks that confront us daily. He has found that some of the most efficient lights, such as high-intensity sodium vapor lights, which deliver light from only the yellow part of the visible spectrum, can actually decrease our visual effectiveness, whereas light with a more balanced spectrum could actually increase productivity by more than 10 percent.[11]

Alternatives

However, there are problems inherent in increasing the spectrum of office lighting. Several of the contaminants emitted by office machines, cleaning materials, and new building materials will react in the presence of UV light to produce smog—precisely as they do outdoors. In one office situation, sunlight-simulating fluorescent lights had to be changed for standard cool-white fluorescents after it was found that the UV light was acting on the hydrocarbon pollutants and generating photochemical smog. Increased ventilation was also called for in this situation.[12]

Continuing research into the properties of light may spawn a new generation of lighting alternatives. Still, the design and creation of office buildings that allow workers access to more natural light is an alternative that must be considered by today's architects.

Tobacco Smoke

Although this subject is treated thoroughly elsewhere in this book, it cannot be overlooked as a critical element of air pollution in the office, especially in buildings with low ventilation. Tobacco smoke is a major source of organic chemical contaminants, odor, carbon monoxide, formaldehyde, particulates, and nicotine, and is a confirmed annoyance to more than 70 percent of the population. Nonsmokers working with smokers have voiced increased job dissatisfaction and tend toward lowered productivity, more days away from work, and a higher likelihood of quitting. The cost to both employees and employers is high.

Building Cleaning Products

Eye irritation and cough

As we have seen in the case of carpet shampoos, products used in the course of building maintenance may be sources of irritating and unhealthy pollutants. In heavily trafficked sites, such as reception areas, stairways, meeting rooms, and school or daycare center classrooms, carpet cleaning may be done as often as every week or two, resulting in a buildup of residue that can permeate indoor air. In the case of a daycare center serving a community college, the carpet was shampooed every three to four months in warm weather and every other week in the winter. After about three years, eye irritation and cough suddenly developed in children and childcare workers while in the classrooms. County officials could find no demonstrable cause of the problem until the childcare coordinator experienced increased symptoms while cleaning up guinea pig droppings from the carpet. Workers began to notice that the children, too, coughed more when they played on the carpet than when they were on the linoleum. When the carpet was steam cleaned, excessive foam was noticed and the source of the problem was found.[13]

Other cleaning products likely to introduce contaminants include dusting sprays, floor wax, furniture polish, and general-purpose cleaning solutions. These emit a wide range of hydrocarbons, formaldehyde, amines, and miscellaneous organic compounds, as noted in previous chapters. Another annoying pollutant is the air freshener frequently used in restrooms. These often add even more annoying, and sometimes sensitizing, odors than those they purport to mask, and they may even have compounds that anesthetize our much-needed sense of smell. Once again, increased ventilation is the healthier solution.

Of course, cleaning and maintenance activities are essential in schools and office buildings, and no one would suggest eliminating them as a cure for the problem of indoor pollution. Proper timing of cleaning procedures, however, could go a long way in improving the air quality for workers. If cleaning is done at the very end of the typical work day, exposures can be reduced almost to zero by the beginning of the following day by means of adequate overnight ventilation.

Humidifiers and Air Conditioners

HVAC systems

As we have previously noted, humidifiers and air conditioners in building-wide HVAC systems are prime breeding grounds for microbes that thrive in their warmth and moisture. Fortunately, epidemics of serious infectious diseases such as Legionnaire's disease (a bacterial pneumonia that attacks several of the body's systems and boasts a 15 percent fatality rate), Pontiac fever (another bacterial pneumonia with flu-like symptoms but a much lower fatality rate), hypersensitivity pneumonitis (caused by a fungal allergen and resulting in serious inflammation deep in the lungs), and humidifier lung (a less virulent respiratory illness probably caused by an amoeba) have generally been rare, but their effects are devastating. Aside from the obvious threat to human health (and life itself), the financial costs are enormous. Affected buildings must be closed down for a complete renovation of the ventilation system, and in some cases, even torn down and replaced.

Water purification

Often, the public is protected from such diseases by chemically treating the water supply in HVAC systems. Periodic chemical treatment is necessary to assure ongoing microbe-free water. Some researchers are experimenting with other means of water purification, such as varying temperatures to create environments hostile to microbial growth. However, additional research is needed in this area.

Ventilation

Evaluation of a building's ventilation system is always a critical part of tracking down the source of building-related illnesses. In the case of infectious diseases, of course, the humidifier, cooling tower, or air conditioner is the obvious prime target. If health investigators fail to uncover evidence of microbial infection, they usually look for a specific chemical or odor brought in from adjacent sources, such as parking garages or air intakes downwind from factories, boiler stacks, or roofing operations. Contaminants originating within the building and circulated by the HVAC system are also investigated. The next step is to examine ductwork for loose fibers or other particulates.

When nothing concrete presents itself as an adequate cause of illness, many investigators are likely to label the problem as "mass psychogenic illness"—that is, mass hysteria. Although such cases undoubtedly do occur (sometimes on the heels of real, confirmed building problems, where additional workers fear they may have been exposed), it may be counterproductive to apply this label too quickly.

Experimental ventilation rates

There is no confirmed fresh-air intake rate that prevents the sick-building syndrome, but in almost all cases, increasing the amount of fresh air drawn into the HVAC system of sealed buildings has alleviated or dispelled all symptoms. Experimental manipulation of ventilation rates may be helpful in finding just the right "mix." Although increases in fresh-air ventilation are at the expense of energy conservation, increased worker productivity and avoidance of more costly problems in the future may be well worth the additional fuel consumption.

Unfortunately, ventilation standards, as suggested by the American Society of Heating, Refrigeration, and Air Conditioning Engineers (ASHRAE), were revised in 1981 back to the lowest ventilation levels since 1830. This was done under the mistaken belief that improved sanitation and personal habits could make up for old-fashioned fresh air (see Table A). Most building operators work only to meet these standards, despite potentially inadequate building maintenance, hazardous building materials, increased cigarette smoking, and pollution-emitting furnishings and office machinery.

Table A: Recommended Commercial Ventilation Rates

Site	Outdoor Air Requirements for Ventilation ASHRAE Standard 62–1981R (cubic feet per minute per person)
Office space	20
Reception areas	15
Conference rooms	20
Supermarkets	15
Theater auditoriums	15
Classrooms	15
Laboratories	20
Libraries	15

ASHRAE = American Society of Heating and Air Conditioning Engineers

A Word to the Worker

Taking power

One of the hardest things to resolve in our mass-produced, mass-oriented, urban industrial society is the feeling of being powerless to control one's own destiny. Workers in large office buildings, surrounded by equipment, smoking co-workers, janitorial policies, sealed windows, fluorescent lighting, partitioned work stations, and the like, are justified in feeling some despair about their helplessness. However, despair is not going to bring about change.

If you, as a worker, feel your office is impinging on your health, take some action. Talk to your co-workers. See how many of them are also dissatisfied with the temperature, humidity, odors, or healthfulness of their nine-to-five environment. With enough backing, pressure can be put on the building's owners or operators to change policies or janitorial supplies or to increase ventilation. If there seems to be a real, shared health problem, don't hesitate to bring it to the attention of authorities such as the Occupation Safety and Health Administration (OSHA) or state departments of labor and public health.

Daily outdoor walks

You can also improve the quality of your own work space. If you are a nonsmoker, try to get the people around you to respect your right to not have to inhale their cigarette smoke. If you are bothered by overhead fluorescent lights, ask for an incandescent desk lamp, or bring in one of your own. Small personal air cleaners or ion generators may make you feel better, although their value is questionable. Most of the clean air they produce may be whisked away by the outtake vent of the building's HVAC system. Taking a walk outdoors during your midday break is another good idea. Even if the weather is cold, the sunshine and fresh air may be just what you need to survive your day at the office.

4 What You Can Do to Lower Indoor Pollution

By now it may look as though everything your house is built and insulated with, and everything you use to clean it, heat it, furnish it, and maintain a family in it with is harmful to your health. You might just as well pack your bags and live in a tent high on some remote mountain peak. But we're not selling mountaintop real estate or advocating that you stay up nights worrying about the fate that awaits you just beyond the next breath.

The problem of indoor pollution is certainly a serious one, but it is also one that the homeowner or apartment dweller can do something about. This is not an issue best left to government legislation or technological intervention. The first steps in improving the quality of the air in your home or apartment lie with you.

Professional help

From reading the checklist in chapter 2, you may have some idea about the number and strength of potential polluters in your household. If these seem especially high and you or members of your family are, in fact, suffering from what seems to be the effects of poor indoor air quality, you may want to get professional help in assessing the concentrations of certain pollutants in your home. There are a number of companies and even government agencies set up to help you do this and even some home test kits with which you can analyze the air in your home (see Appendix for source list).

Home test kits

Because an EPA study of 10 states revealed that more than 20 percent of the homes tested had radon levels elevated enough to pose a long-term health risk, the federal government recommends that all homes (and schools) conduct at least a preliminary radon screening, for which a simple test kit is now widely available. In the case of other pollutants, however, measurements are most appropriate when either health symptoms are noticed, major sources are known to be present (such as in new mobile or pre-fab homes) or extremely low ventilation is known to be the case (such as in highly weatherized housing).

The home test kits available are generally unobtrusive and easy to use. For gases such as nitrogen dioxide or formaldehyde, for example, small vials containing nonpoisonous chemical adsorbent membranes are hung in the room to be tested, away from walls or drafts, and left in place for five to seven days, during which time the chemically impregnated filter traps the designated gas at a constant rate. The vials are then capped and sent back to a laboratory for analysis. The results offer a fairly accurate idea of the average level of pollutant during the test period. The concentrations are given in parts per million. A concentration of one part per million means that there is one part of, say, formaldehyde gas, for each million parts of air. The company returns the results with sufficient information for the average consumer to determine if his or her concentrations are significant, and suggests ways of improving the air quality where levels are high enough to impair health.

Industrial-type monitors (often called "personal monitors") are also available. With these, a person within the home wears the monitor as a badge on his or her collar (near the breathing zone) for a specified length of time. The mechanics of the monitor and the test results are analyzed in a manner similar to the procedures used for the room monitors.

Radon is generally monitored in the home in one of two ways. The simplest test, a short-term monitor, is typically a canister of activated charcoal granules that trap radon gas. This canister is opened and placed in the basement (or lowest floor of the house), preferably with any doors or windows closed. After a period of time (generally three to seven days) the canister is resealed and sent to the lab for analysis. The cost for the canister, postage and analysis is generally around $20.

If a radon level of more than about 3 picoCuries per liter (3pCi/l) but less than 20 pCi/l is found a second test should be conducted *before* engaging in any abatement activities. This is done to rule out either a testing error or simply an unusual fluctuation of levels. This second test should be made with a long-term detector, commonly called an "alpha track device," several of which are placed in the general living areas of the house.

This technique consists of exposing a plastic detector to the atmosphere. Alpha particles from the air or attached to dust particles bombard the plastic and cause radiation damage tracks that are subsequently revealed by caustic etching and counting under a microscope. The plastic detectors are generally located on a few surfaces in a room or rooms within the home and at suspected points of radon entry.

The length of exposure can be varied, but is typically three to twelve months. The detectors are then sent back to the lab for analysis. The cost for this procedure is about $25. The EPA has established a Radon/Radon Progeny Measurement Proficiency Evaluation and Quality Assurance Program to evaluate the companies offering radon measurement services, and both they and state environmental protection departments can provide a list of approved testing and abatement companies.

If screening measurements over 20 pCi/l are reported, EPA recommends a more intensive follow-up measurement with state or local government assistance, because such concentrations could cause a significant increase in health risk. In the case of high concentrations of other pollutants, government and private laboratories can be notified and requested to step in with more complex (and costly) testing equipment that yields immediate results. Such equipment is rarely offered for home use.

Four approaches

Even without outside intervention, however, you can begin to clean up your personal environment. There are four basic approaches to a healthier indoor climate: Removing or modifying the source of pollution (and replacing it with a low-pollution substitute); increasing ventilation to remove and dilute contaminants; isolating the contaminant or its source from the living space; and air cleaning. Each of these approaches has certain advantages, limitations and, sometimes, costs. They are not meant to be mutually exclusive, either. In fact, they work best in combinations, such as source modification plus ventilation or air cleaning.

Source Removal

Making substitutions

By far the most effective means of pollution control is source removal. After all, no polluter equals no pollution. Many sources lend themselves to this response: You can prohibit smoking in your home, eliminate gas ranges and portable space heaters, and so on. Along with source elimination, however, you will very likely have to make substitutions in order to remain comfortable and satisfied with your standard of living. In some cases where complete source removal is not practical or desirable, modifications may be possible to at least lower, if not eliminate, the pollution emissions from a given source.

Table A lists items that can be eliminated, and suggests modifications or substitutions.

Ventilation

As you can see, not all your pollution sources can simply be eliminated. Many are inherent in the

Table A: Indoor Pollution Sources for Elimination and Modification

Sources for Elimination	How to Modify	What to Substitute
Gas stoves, dryers	Vent to outdoors Buy pilotless appliances	Electric appliances
Cool-mist humidifiers	Disassemble and scrub parts weekly	Ultrasonic humidifiers
Kerosene and gas space heaters		Electric space heaters
Asbestos pipe insulation (only by a professional)	Encase securely	Non-asbestos pipe insulation
Urea-formaldehyde foam insulation	Seal the interior walls carefully and completely to keep formaldehyde fumes from penetrating	Cellulose insulation Glass fiber insulation
Cigarette, cigar smoking		
Spent or half-used and unwanted aerosol spray cans, e.g., cosmetic sprays, paints, solvents, pesticides, etc.		

structure or location of your home or are too high a priority in your chosen lifestyle. This is where ventilation becomes important—to disperse the pollutants you cannot eliminate. Ventilation sounds easy—just open the window and clear the air, right? Maybe in clear and balmy weather, but not when you are losing all the nicely heated or cooled air you have spent so much money to create. Here, ventilation needs to be a bit more sophisticated.

The purpose of ventilation as we will discuss it is twofold: To remove the offending pollutant as effectively as possible at its source and to add sufficient fresh air to the rest of the house so that the remaining pollution is diluted to a safer concentration. Before the recent energy crunch, this last process was adequately accomplished by the air leakage, or infiltration through a house's walls and around windows and doors. The added insulation, vapor barriers, caulking, and sealing that constitute energy efficient housing virtually eliminate this fresh air infiltration and must be compensated for by other means.

Exhaust ventilation

In some cases, exhaust ventilation at the source is the best way to go. For example, gas stoves can be fitted with range hoods and exhaust fans that draw in the air and effluent over the cooking surface and blow them outdoors. A word of caution: Many homes are equipped with range hoods using charcoal filters that clean the air and then vent it back into the room. These filters may be useful in removing grease, odors, and some large molecules (at least until they become dirty), but they are totally ineffective in controlling carbon monoxide, nitrogen dioxide, and the other previously mentioned pollutants generated by gas cooking. If anything, they help disperse these throughout the rest of the house. Therefore, they should not be mistaken for adequate ventilation systems.

If a proper exhaust hood is not in your budget right now, and neither is a new electric stove, a low-tech response might be to open a window near the stove and fit it with a fan blowing out. This will at least eliminate some of the combustion products, although it is clearly not energy efficient in temperature extremes.

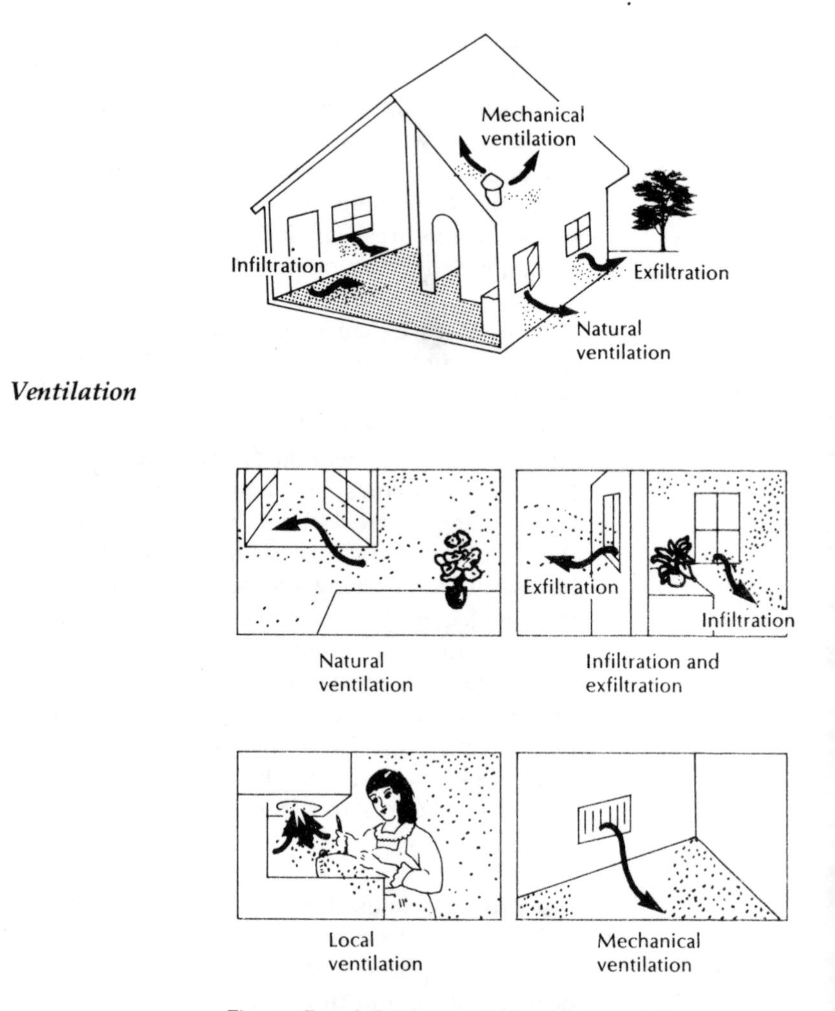

Figures B and C: In natural air exchange, differences in air pressure between indoors and outdoors cause outside air to enter through open vents, windows, doors, and cracks in the building's superstructure. Mechanical ventilation uses fans to force air exchange.

The next step is to provide a steady influx of fresh air that does not require a disproportionate amount of energy to heat or cool to your needs. Simple air duct fans or exhaust fans in attic windows or air vents will draw out stale air and increase the rate at which it is replaced with fresh air, but this influx of fresh outdoor air will have to be heated or cooled at great expense when indoor-outdoor temperatures vary by 20 degrees or more. In addition, the presence of air exhaust fans removing air from a too-tight house could encourage an influx of contaminated air from the furnace system.

Heat exchangers

Fortunately, technology provides a safer, more energy-saving answer in the form of something called an air-to-air heat exchanger. This is a unit that pulls warm, moist, stale air from the house and transfers the heat in that air to fresh, cold air being pulled into the house. (It can also work the opposite way in hot weather, precooling the air from outdoors by allowing it to transfer its heat to the stale indoor air being exhausted.) Such heat exchangers are now in common use in tightly built (highly weatherproofed) houses in Scandinavian countries.

Although the heat is exchanged between the two airstreams, the indoor and outdoor air do not mingle, and pollutants are not transferred from one to the other. Tests show that with the exchangers, 50 percent to 70 percent of the available heat is transferred, taking a great deal of the burden off the house's heating or cooling system.

There are basically two types of residential heat exchangers: Central systems that use the building air ducts to collect and disseminate air, and small units that resemble ordinary air conditioners and can be mounted in a window or wall sleeve. The central systems, which are generally recommended for new construction, do work more efficiently, whereas the small units are less expensive and simpler to retrofit in an existing home.

Fixed-plate exchangers

Today's commercially available heat exchangers recover heat in one of three ways. The most conventional design, the fixed-plate type, has no moving parts other than the fans. As one fan draws air from within the house the other draws air from outside. The core has a large surface area of narrow, thin-walled passages made of heat-conductive material through which the air travels. As the two airstreams slowly move through the passageways, passing one another, the cooler airstream absorbs heat from the warmer one, in accordance with the laws of physics, and the job is done. The passages can run parallel with one another (counterflow) or at right angles (crossflow).

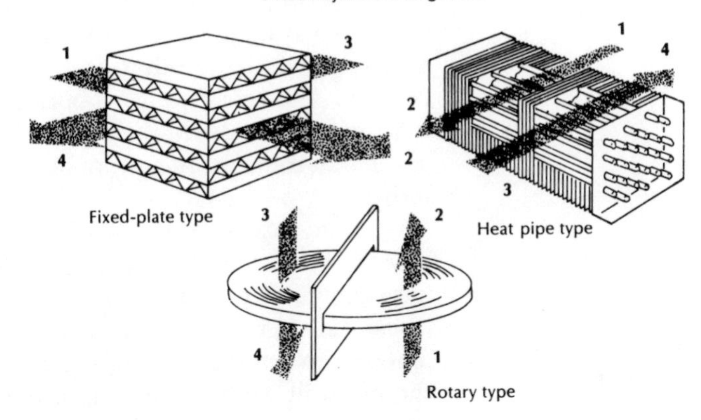

Figure D: A heat-exchange element is the heart of a heat-recovery ventilator. Fresh outdoor air (1) is warmed as it passes through the exchanger and enters the house (2). Stale indoor air (3) leaving the house is cooled as it transfers heat to the exchanger and is vented outside (4). In the fixed-plate type, heat is transferred through plastic, metal, or paper partitions. The turning wheel of the rotary type picks up heat as it passes through the warm air path and surrenders the heat to the cold air stream half a rotation later. Liquid refrigerant in the pipes of the heat pipe type evaporates at the warm end and condenses at the cold end, transferring heat to the cold air. (Courtesy Consumer Reports.)

Rotary exchangers

Another design, the rotary type, uses a wheel that rotates across both air passages to transfer the heat. The porous material of the wheel allows air to pass through it. The warm air heats the wheel which, in turn, heats the cool incoming air.

The third type, the heat pipe, consists of an array

of refrigerant-filled pipes lying across two air passageways. One end of the pipe is in the warm airstream, and the other end is in the cold airstream. The refrigerant evaporates at the warm end and condenses at the cold end, repeating in a cycle that transfers heat from one airstream to the other.[1]

The fans driving the airstreams are the major consumers of power in all of these devices, and these are generally small and use minimal energy, to avoid offsetting energy savings from a home built to be energy efficient. Other basic features of a heat exchanger might include a timer and a humidistat to ensure the proper amount of ventilation, and a defrosting mechanism to prevent water vapor condensation from freezing in extremely cold weather. Some heat exchangers are equipped to recover humidity as well as heat. This minimizes the problem of condensation in the heat vent, and adds moisture to the air in your home.

From an economic standpoint, heat exchangers are a sensible solution in very tight houses in climates with a tendency toward temperature extremes, especially in areas where serious indoor pollution problems such as radon gas are likely. Several manufacturers are now offering heat exchangers; a partial list of manufacturers and/or distributors appears in the Appendix.

Houses with adequate air exchange rates (0.5 to 1.5 air changes per hour is typical for American homes and apartments) and mild to average pollution source strengths may benefit more from other means of air cleaning or pollution control. Table B will help you decide whether your house is a candidate for a heat exchanger.

Table B: Diagnosing Your Need for a Heat Exchanger

You May Need a Heat Exchanger if Your House:

A. has a vapor barrier (e.g. Tyvek);
B. is damp in the winter;
C. tends to collect and keep odors;
D. has double- or triple-glazed windows;
E. cannot be equipped with local exhaust ventilation over stoves or in bathrooms;
F. is situated in an area with high radon-content bedrock;
G. faces extremely cold winter temperatures.

Isolation

It is best to deal with certain sources of pollution in your home by isolating them and preventing their emissions from ever entering the indoor environment. Radon gas emanating from building materials and underlying or surrounding soil is a primary example.

Radon

Cracks and holes in foundation, basement walls, and concrete slabs can be filled and sealed with polymeric caulking to prevent radon-laden air from penetrating into the house, and crawl spaces can be specially ventilated to prevent the buildup and further penetration of any radon that travels that far. In addition, building materials that are judged to have high emanation rates can be sealed with epoxy resin paints, polymeric sealants, or polyethylene or polyamide film (vapor barrier) to prevent the escape of the radon. These techniques can substantially reduce radon concentration indoors in the short term, although their long-term effectiveness is not yet known.

Recent research has shown that far more effective techniques can be used permanently to lower indoor radon concentrations. One particularly effective technique for rapidly lowering excessively high radon levels is called subslab suction. It involves retrofitting pipes underneath a house and attaching them to a fan that draws gases out from under the house and releases them into the open air. The cost of such a project can be as high as $1,500 and it takes about one day for installation.

The best way to attack radon, however, is from the beginning. At the most basic level, builders should avoid practices that make it difficult to mitigate radon problems after construction. This can be done simply by including porous materials under the slab, capping the tops of hollow foundation walls and avoiding water drainage systems that leave large openings at the top of the slab. While these particular measures do little to reduce radon levels, they allow any future mitigation techniques to be retrofitted for the least amount of money.

Builders can also install "passive" subslab depressurization systems that do not use a fan, but are powered by warm air that rises in a special stack that is vented at the roof. This sort of system may add only about $250 to normal construction costs, but it may be easily converted to an "active" system (if necessary) by the addition of a fan in the stack.

New housing construction may soon benefit from radon abatement guidelines written by the EPA and designed to make sure that indoor radon exposure is no greater than ambient conditions outside the home.

The 1990 annual convention of the National Association of Home Builders devoted time to educating builders about their role in lowering indoor radon levels. Two basic options in new construction currently exist. Both include a gas permeable subslab layer, plastic sheeting to prevent radon gas infiltration below the foundation, sealing of plumbing and ductwork openings, joints and access spaces, non-closeable vents in storage areas, sealing of below-grade penetrations, and everything but a motor and fan for a subslab ventilation system, exhausting through the roof.

Option 1, the most thorough approach, requires completing the forced subslab ventilation system, but would then relieve builders of any further cost or responsibility for radon risks. Under option 2, builders could test radon levels for a year after the building is completed. If the tested level is above 2 pCi/l, they must complete the subslab ventilation system. Although these guidelines do not represent enforceable standards, builders (especially those in known radon risk areas) will likely find them preferable to the threat of future lawsuits over buildings in which radon levels reach health-threatening levels.

GAC tank

Radon gas dissolved in well water can be effectively removed using a granular activated carbon (GAC) adsorption unit to treat the water before it enters the house. A GAC unit is a tank containing a bed of activated carbon that is capable of removing low concentrations of impurities. The unit is usually installed in the main water supply line after the pressure tank. The water supply passes through the GAC where the radon is adsorbed onto the activated carbon surface and then continues on into the house. Periodic backflushing of the unit cleans the filter and maintains efficient radon removal. Since a small amount of the activated carbon is lost during backflushing, GAC units will need replacing, although probably not more than every 10 years. Because they concentrate the radon and its progeny, GAC units are themselves considered sources of low-level radiation (well below federal guidelines), and for safety's sake should be located in the cellar or basement.[2]

Formaldehyde

Formaldehyde outgassing from UF foam insulation can also be at least partially controlled by means of vapor barriers, or even vinyl wallpaper or low-permeability paint applied on interior walls. The same is true for the formaldehyde in particleboard and plywood, which can be coated with shellac, varnish, polymeric coatings, or other low-diffusion barriers. These barriers act to contain the outgassed formaldehyde, which is seemingly reabsorbed by its source rather than released into the home.

Air Cleaners

That portion of indoor air pollution that cannot be sealed off, eliminated, exhausted, or diluted to a safe enough level can often be removed from the indoor environment by using special types of air cleaners. There are several types of air cleaners, and they are especially good for alleviating the problems of tobacco smoke, radon gas, dust, and other particulates. It should be noted, however, that within the limits of today's technology, no single type of air cleaning equipment can control gases, vapors and particles.

The air cleaners available for residential purposes use the principles of filtration, adsorption, and electrostatic precipitation, either individually or in tandem, to do their job, and it helps to understand how each of these works in order to choose the best air cleaner for your needs.

Filtration

Filters made of charcoal, glass fibers, vegetable matter, synthetic materials, or even paper may be used to remove particles from the air mechanically when air is gathered from a room by fan and pushed or pulled through the filter material before being blown back into the room. This method's effectiveness depends to a large extent on the volume of air needing to be cleaned, the size of the unit, the rate of airflow, and the density of the filter itself. It also depends on the types of particles that need to be removed. For example, pollen or lint, which are relatively large particles, are easily trapped by most filters. Central air conditioning systems and most residential furnaces come equipped with a filter capable of this.

HEPA filters

Smaller particles, such as those in cigarette smoke, are more difficult to trap and require more specialized filter material, such as medium-efficiency filters and the recently developed high-efficiency particulate air (HEPA) filters (also known as "extended surface filters") that can remove almost all particles larger than 0.3 microns, which includes bacteria and spores but not viruses. These HEPA filters are most often used in industrial applications, nuclear reactor facilities, and the surgical suites and isolation units of hospitals. Small HEPA filters have recently been developed for use by highly allergic or asthmatic patients, to create a "clean space" around their heads as they sleep. Their effectiveness requires more study. Another type of filter, called electrets, carries an electric charge (negative or positive) to attract particles electromagnetically.

All filters need periodic cleaning and/or replacement in order to remain effective. Most of the better filtration devices on the market today allow for easy removal of the filter, and this should be one of the criteria for selecting an air cleaner for your home.

Adsorption

Adsorbents are similar to filters in that they are porous materials with a large surface area and tiny pores through which gases can pass. But whereas filters simply trap the larger particles, adsorbents react with the molecules, causing them to cling to the walls of the pores. Three common adsorbents are activated charcoal, activated alumina, and silica gel, and these may be additionally impregnated with such materials as sodium sulfite or potassium permanganate to enhance their efficiency. Adsorbents are better at removing gases than are filters, although even their effectiveness is limited to gases with large molecules such as formaldehyde and ammonia. Generally, adsorbents are used in conjunction with fan/filter devices; and as with conventional filters, adsorbents need to be replaced periodically in order to remain effective.

Electrostatic precipitation

In its absolutely simplest sense, electrostatic air cleaners work by charging airborne particles with either a negative or positive electrical charge. Once the particles have such a charge, they act magnetically, gravitating either to special collector plates within the air cleaner or to objects in the room itself (such as walls, furniture, carpeting) that carry the opposite charge. Graphically, you might even think of the particles as iron filings being attracted to strong magnets around the room.

An intake fan blows air from the room through a highly charged electrical field inside the air cleaning unit, and the particles in the air receive a like charge (usually positive). The air then proceeds through a second electrical field between a series of oppositely charged plates to which they attach themselves. The technique is reasonably effective against dust, smoke particles, and some allergens, as long as the plates are kept clean, usually by washing with soap and water.

One of the drawbacks with some electrostatic air cleaners is that they do not contain collection plates or the collectors become ineffective. This causes the walls and other surfaces in the room to become soiled as the charged particles adhere to them. When they are cleaned or dusted, the particles may resuspend themselves in the surrounding air. Washing the walls periodically with a damp cloth can partially solve the problem but this is time consuming and difficult, and therefore no real answer. Another complaint about a few of the electrostatic models is that they produce ozone, another environmental pollutant. Manufacturers have been sensitive to these complaints and have attempted to rectify the problem by developing better models.

Negative-ion generator

Another sort of electrostatic air cleaner, with reputed additional side benefits, is the negative-ion generator, a machine that sprays a continuous fountain of negatively charged ions into the air. When these ions collide with particles, they transfer their charge, again causing the charged particles to be attracted to oppositely charged surfaces. Again, there is the problem of soiled walls, although some makers claim to have circumvented that by adding intake fans and collector plates. The extra health benefits are said to come from the presence of a large number of negative ions in the air—a situation that some researchers believe is conducive to a general feeling of well-being, increased mental and physical energy, and relief from some of the symptoms of allergies, asthma, and chronic headaches. There are also claims that an abundance of negative ions speeds up the healing process and reduces pain for burn victims and sufferers of peptic ulcers and other chronic ailments. Although further research must be done to substantiate these claims, the air-cleaning properties are well documented.

Effectiveness of air cleaners

New Shelter magazine tested some 20 portable air cleaners, including fan/filter units and electrostatic models. The tests were conducted in a small (1,200 cubic foot) room with a cigarette-smoking machine loaded with two unfiltered Lucky Strikes. After the cigarettes were smoked, the air cleaner to be tested was switched on for four hours or until all the smoke was removed, whichever came first. Monitoring instruments automatically took continuous samples of air in the room and recorded the concentration of smoke particles on chart paper, showing how fast the smoke was dissipated. In addition to testing air cleaners' effectiveness in smoke removal, units were tested for noise and ozone emission.

According to the results, the fan/filter units showed only marginal improvement over using no air

Air cleaners

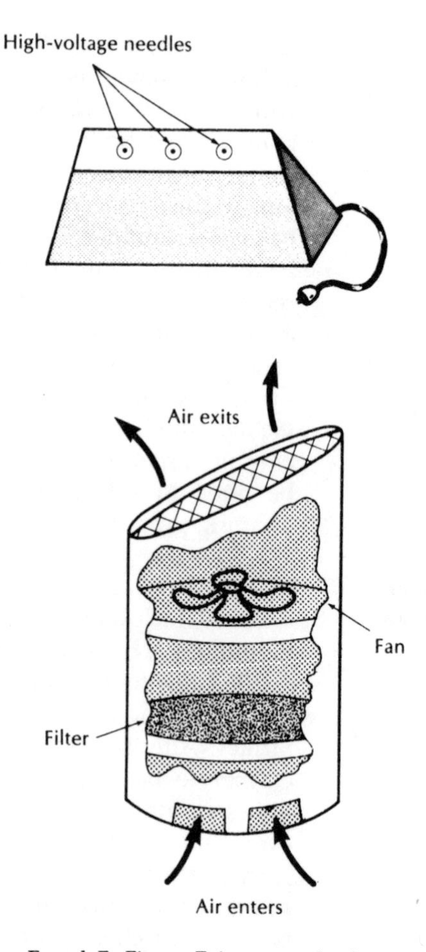

Figures E and F: Figure E is a negative-ion generator. In Figure F, a fan/filter air cleaner, air is sucked through a filter medium that traps large particles.

cleaner at all (out of 12 units, nine of them removed only about 24 percent of the smoke in four hours, whereas when no air cleaner at all was used, about 17 percent of the smoke settled out of its own accord).

The five negative-ion generators did much better: Four of them removed over 96 percent of the smoke in four hours or less, the fifth removed 70.4 percent in four hours. None of them released any significant amounts of ozone, according to the magazine. Another machine—a larger hybrid unit combining a fan/filter air cleaner and a negative-ion generator—cleared all smoke from the room in 1 hour, 46 minutes. In addition, the magazine tested a high-capacity air cleaner with a prefilter plus an extended surface (HEPA) filter, and that unit cleared the air completely in 2

Two-step precipitators

Electrostatic Precipitator

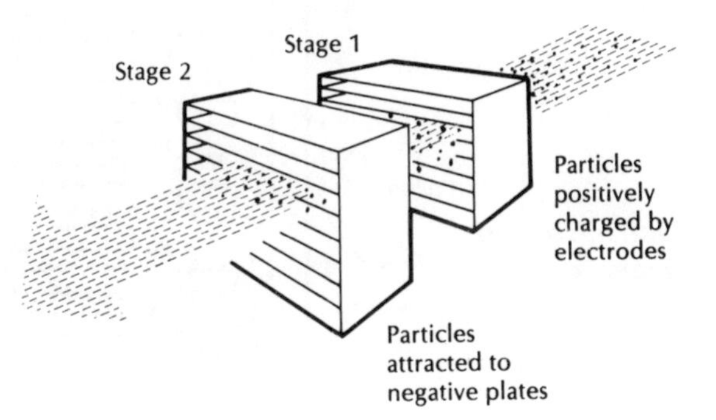

Figure G: In the two-step electrostatic precipitator pictured above, air is pulled through the first stage where particles receive an electrical charge from a high-energy field. In the second stage these charged particles are attracted to oppositely charged plates before the air is blown back into the room.

hours, 41 minutes. The fan/filter units were also judged to be annoyingly noisy in operation, whereas the ion generators were almost silent. Some of the latter were equipped with collector plates to avoid soiling room surfaces.

New Shelter also tested a simple, portable, three-speed fan operating within the room, and they found that simply by knocking the smoke particles against the walls, it managed to clear nearly 54 percent of the smoke in the four-hour period, when the door was closed. With the door open,

Multi-drive air cleaners

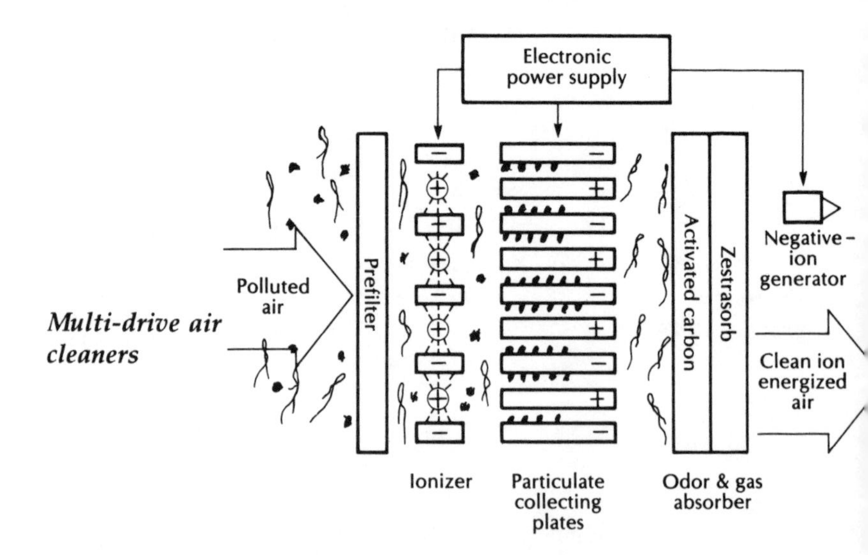

Figure H: An air cleaner that makes use of several principles of air cleaning including filtration, electrostatic precipitation, adsorption, and negative ion generation. (Courtesy Zestron Products.)

they noted, the fan removed over 25 percent of the smoke in just 40 seconds and virtually all the smoke in under six minutes. They concluded that although the electronic air cleaners are clearly superior at removing particles from a closed room, the most cost-effective way of cleaning the air during appropriate weather may be to open the windows or doors and turn on a fan.[3]

Tests

Researchers at Lawrence Berkeley Laboratory also tested 10 air cleaners in a 1,241-cubic-foot test space with particles generated by a smoking machine. Their results were surprisingly different from those of *New Shelter*, although again, the small fan/filter devices were the least effective. In these tests, however, the best performers were the two extended surface (HEPA) filters and the two electrostatic precipitators. The negative-ion generators were found to be far less effective than they were in the *New Shelter* tests. One (the ISI Orbit, for residential use) showed almost no effect in removing the smoke particles, and the other (the Zestron Z-1500, designed primarily for commerical use) performed better than the fan/filters but less effectively than the HEPA filters.[4]

Caveats and guidelines

No air cleaning system is currently available that will entirely remove all pollution from indoor air, and controversy exists about their effectiveness in handling such large particles as pollen, house dust allergens, some molds and animal dander. Source removal and increased ventilation remain the primary means of improving indoor air quality in most cases. Nevertheless, under the right conditions air cleaners can effectively decrease the concentration of suspended particles and even a portion of the gaseous element of indoor pollution, thus reducing the health hazards.

According to an article in the December 18, 1989 issue of *Air Conditioning, Heating & Refrigeration News,* the newest generation electronic air cleaners can arrest many more microscopic particles (as small as 0.01 microns) boosting efficiency to levels 20 times greater than conventional filtration cleaners. These new cleaners use electricity to attract up to 95 percent of all airborne particles. Energy efficient air cleaners use about as much electricity as a 40 watt lightbulb.

Although the federal government has not published any guidelines or standards for use in determining how well an air cleaner works, a private standard-setting trade association—The Association of Home Appliance Manufacturers (AHAM)—has developed an American National Standards Institute (ANSI) approved standard for portable air cleaners. The standard is expressed in the clean air delivery rate (CADR) for each of three particle types in indoor air: tobacco smoke, dust, and pollen. The CADR is a measure of the number of cubic feet per minute of air it cleans of a specific material. For example, if an air cleaner has a CADR of 150 for smoke particles, it would reduce smoke particle levels to the same concentration as would be achieved by adding 150 cubic feet of clean air each minute.

Unfortunately, only a limited number of air cleaners have been certified under the AHAM program. A complete listing of these and their CADR's can be obtained by sending a stamped, self-addressed envelope to:

Association of Home Appliance Manufacturers
Air Cleaner Certification Program
20 North Wacker Drive
Chicago, IL 60606

It should also be remembered that any air cleaner will only perform up to its standards if it is properly installed and maintained. This means that filters and sorbents must be cleaned or replaced and plates or charged media of electronic air cleaners must be cleaned regularly. When cleaning filters, excessive movement or air currents should be avoided to prevent particles from being redistributed in the air.

Prevention

Household products

As you can see, there are many options for you to take advantage of, once you decide to clean up your indoor environment. As with so many things, however, the best cure for the problem of indoor pollution may be its prevention. Wherever possible, you should work to safeguard the quality of indoor air by making sound and informed choices on such basics as site selection, building materials and design, insulation materials, decorative devices, and products for household use.

Evaluations

This book does not purport to be a guide to the design and building of the pollution-free home, and of course, few of us are ever offered the option of designing our dream house from scratch. But even when buying a new house or changing apartments we can evaluate our new home better if we take into account certain criteria. Table C lists key factors in building construction, materials, and design that may affect indoor air quality, and Table D provides a guide to the selection of household products.

Table C: Building Design and Construction Factors Affecting Indoor Air Quality

Factor	Potential Effects
Site characteristics	The geology of the area may be a key factor in determining the potential radon infiltration into the home. Foundations dug into former mine residue or waste dumps can provide pathways for radon residing in the mine tailings. Granitic rock is another potential source of radon where it has a high radium content. Chemical waste dump sites and other landfill sites can also be sources for radioactive and chemical pollutants. In some cases, the offending soil can be removed and replaced with unpolluted earth into which the foundation can be dug.
Water source	Well water found in certain areas of granitic rock, high in radium, can have a high dissolved radon content that can be released inside the home during activities such as showering, laundry washing, toilet flushing, etc. This is virtually never a problem with water from a reservoir, however.
Outdoor air quality	Residences built near a great number of factories, refineries, mines, major streets, or highways may be subject to a great deal of indoor pollution from infiltration by outdoor air. This is one case where a tighter, more weatherproof residence could actually be lower in pollution concentration than a leaky, airy one would be.
Building materials (exterior)	Brick, stone, concrete, and gypsum board may have a high radon content (which can be measured).
	Asbestos tile facades may allow asbestos fibers to enter as they weather and fibers loosen from their bonding.

Plastics, used in place of glass windows, may degrade slowly and release organic and nonorganic chemical vapors.

Housing

Tarpaper used on exterior walls and roof can also outgas organic vapors that can enter the house by air exchange. The new, high-tech building wraps (vapor barriers), such as Du Pont's Tyvek, reduce air exchange significantly and may cause moisture to collect between exterior and interior walls, leading to chemical leaching from materials, mildew and rot, and increased formaldehyde outgassing.

Solvents from paints and sealants evaporate to contaminate indoor air.

Caulking and weatherstripping reduce air exchange, and caulking may be a source of contaminants by outgassing during the curing process.

Building materials (interior)

Interior grade plywood and particleboard and most paneling are bonded with formaldehyde resins. Because this grade is not waterproof, moisture absorbed into these materials will promote outgassing of the formaldehyde.

Insulation

Any form of insulation is designed to reduce air exchange which also slows diffusion of contaminants into or out of the residence.

Insulation in the floor above a crawl space can slow diffusion and infiltration of radon.

Urea-formaldehyde foam insulation can outgas significant amounts of formaldehyde into the residence, especially when new, unless it is well-sealed from the interior of the house.

Shredded cellulose insulation can settle and attract moisture, making it a breeding ground for mold and mildew that can cause surrounding material to deteriorate.

Asbestos pipe insulation and air-duct sealant can release asbestos fibers as it ages. Troweled-on asbestos used for soundproofing or decorative effect can also release fibers when cleaned or disturbed by loud vibrations.

Wall-to-wall carpeting placed on a rubber mat directly over a concrete slab floor provides thermal and noise insulation, but carpeting may become saturated with gas, liquid, and particle impurities and become bad-smelling as well as a source of contamination.

Heating and cooling systems

Systems that affect interior airflows (such as hot and cool air duct systems) can reduce indoor particulate concentrations when appropriate filters are in place and properly maintained. However, they can also help disperse pollutants throughout the house (e.g. combustion products from boilers, stoves, and ovens; pesticides used in one area of the house or in the basement; and asbestos fibers from pipe insulation in the basement).

All-electric heating and hot water systems emit substantially fewer pollutants, although they may be less efficient users of energy.

Inside the house

Gas and oil heating and hot-water systems can boost the indoor concentration of carbon monoxide, nitrogen oxides, and sulfur dioxide even under proper working conditions. When they are improperly maintained or when rust, cracks, or leaks develop in the flue pipe, potentially lethal amounts of these gases can escape into your home. A chimney blocked by debris is also hazardous. Soot or carbon deposits on the heating system often indicate the need to have the burner adjusted. Proper air flow into the burner can be tested by holding a hand over the air diverter while the burner is on. If the system is working properly, you will feel suction. If you feel warm air, the flue or chimney may be blocked.

Fireplaces

Fireplaces are major producers of combustion pollutants. However, fireplaces that have air intakes allowing them to use outside air for combustion will disperse far fewer of these pollutants into the room.

Solar heating systems

The rock thermal storage medium (also known as a trombe wall) may be a source of radon or prime location for microbial growth.

Ventilation systems

Although vents and hoods are meant to exhaust contaminants to the outside, air intakes located too close to these will reintroduce contaminated air into the house.

Building design

Taller buildings have greater exposure to increased "stack effect," drawing more fresh air inside and increasing infiltration. "Open" floor plans (fewer walls and doors) affect air flow and may help spread pollutants throughout the building.

Table D: A Guide to Household Product Selection

Please note that virtually all the products listed here are generally regarded as safe for their intended use, according to directions, in a well-ventilated home. However, key ingredients of these products are toxic or noxious when ingested or inhaled in sufficient concentrations. This chart has been included to guide the consumer in the effort to minimize pollutants in the home in order to maintain a healthful indoor environment.

Product Type	Some Undesirable Ingredients	Typical Brands Containing one or more of These Ingredients	Alternative Brands & Home Recipes
Scouring powder	chlorine	Comet Ajax (most local and generic brands)	Bon Ami plain table salt Soft Scrub
General purpose cleaners	ammonia petroleum distillates (organic solvents) ethylenediamine ethyleneglycol	Lestoil Top Job HSC BeautiFi Lysol Deodorizing Cleanser Ammonia Fresh Ajax	Murphy's Oil Soap Mr. Clean LOC (from Amway) Basic H (Shaklee) Basic I (Shaklee) Spic and Span Pine vinegar mixed with water baking soda on a damp sponge (rinse with water and polish to shine) most soap pads (e.g. SOS pads)
Window cleaners	ammonia petroleum distillates (organic solvents)	Windex most local and generic brands Glass wax	*Recipe:* 2 Tbs. vinegar to 1 qt. water: Apply with wadded-up newspaper
Disinfectants/ Bathroom cleaners	phenol; cresol; aerosol propellents n-ethylmorpholine ethylenediamine sodium hydroxide (lye) sodium hypo-	Lysol Disinfectant Spray Pine Action Lysol West Pine Wilbert Fresh Pine Dow Bathroom Cleaner Disinfectant	Arm & Hammer Washing Soda Super 80 Pine Disinfectant Pine Sol Real Pine baking soda on a damp sponge (rinse and polish to shine

	chlorite (chlorine bleach)	Dow Tough Act Bathroom Cleaner Spray Lysol Basin/Tub/Tile Cleaner Tilex Tackle Cleaner/ Disinfectant	tiles)
Toilet bowl cleaners	hydrogen chloride other eye and mucous membrane irritants	Lysol Liquid Toilet Bowl Cleaner Purex Sno Bol Toilet Bowl Cleaner Vanish Scrub Free Heavy Duty Toilet Bowl Cleaner	See Scouring powder
Furniture polish/ wood floor cleaners/dusting sprays	spray propellents petroleum-based solvents benzene compounds 1,1,1 trichloroethane	Pledge (spray) Lemon Pledge Aerosol Old English (spray) Wood Preen Holloway House Lemon Oil Furniture Polish Favor Lemon Furniture Polish (spray) Scotts Liquid Gold Duster Plus	pure lemon oil (not spray) Scand-Oil *Recipe #1:* 1/2 c. vinegar 1 c. linseed oil 1/2 c. rubbing alcohol Shake well before each use. *Recipe #2:* Cigarette ash on a moistened cloth dispels cup rings on polished wood. Rub vigorously, then polish.
Spot remover/ upholstery cleaner	organic solvents benzene; perchloroethylene; naphtha; trichloroethylene; toluene; 2-butoxy ethanol; aerosol propellents 1,1,1 trichloroethane	K 2-R Spot Remover Scotchgard Upholstery Cleaner Carbona	Fuller's earth sprinkled on the spot for 1/4 to 2 hours.
Rug cleaner	harsh detergents aerosol propellents "cis" & "trans"	Woolite Self-Cleaning Rug Cleaner Glamorene Spray'n Vac Sargeant's Rug Patrol	Procter & Gamble's Orvus (granules or paste) Airwick Carpet Fresh baking soda mixed with water

Oven cleaner	aerosol propellents sodium hydroxide (lye)	Easy-Off Oven Spray Mr. Muscle Oven Cleaner	SOS No Fume Oven Cleaner (pad form; contains lye not released and not hazardous)
Laundry bleach	sodium hypochlorite (chlorine bleach)	Clorox most generic and local brands	Clorox 2 Oxydol Vivid
Air freshener	propylene glycol morpholine ethanol; cresol aerosol propellents	all spray products	
Insecticides	aerosol propellents (especially "room foggers") carbaryl (Sevin) trichloroethane petroleum distillates dichlorovinyl dimethyl phosphate (Dichlorvos)	King Spray d-Con (spray) Black Flag (spray) Raid (spray) Black Jack (spray) Formula 707 (spray) Black Flag Large Area Automatic Room Fogger Raid Fumigator Raid Fogger Safeguard Roach and Insect Killer Automatic Continuous Spray Texise No-Pest Strip Insecticide d-Con four-gone Automatic Room Fogger Ortho Hi-Power Indoor Insect Fogger	TAT Roach Traps TAT Ant Traps Roach Prufe Roach Motels (Black Flag) King Spray Roach Traps ordinary boric acid (Do NOT use in places accessible to children and pets.) Combat Roach Wrecker (Safeguard) Raid Waterbug and Roach Traps Black Jack Waterbug and Roach Traps Black Jack Bug House Enuf (roach powder) Dro (roach powder)
Houseplant insecticides	systemic insecticides e.g. carbaryl (Sevin)	Isotox	Green Ban Wash plants often in soapy water; rinse well

Ingredients taken from labels as of July, 1986. Many household products do not list their ingredients on their labels. Where the names of such products appear in this chart it is due to their aerosol spray form.

5 Should the Government Regulate Indoor Air Quality?

■ A family on Long Island, New York was forced to vacate their home shortly after having it treated with the pesticide chlordane because concentrations of the chemical remained at health-threatening levels.

Case histories

■ After a New Jersey family went to the trouble and expense of having urea-formaldehyde foam insulation installed in their home, members of the family began experiencing chronic bouts of headache, nausea, irritability, nosebleeds, and respiratory problems. When the problems were attributed to formaldehyde, the family had to spend an additional $20,000 to have the insulation removed.

■ Hundreds of families living in housing developments built on the sites of former industrial landfills have experienced health symptoms ranging from chronic respiratory illnesses to multiple miscarriages in pregnancy and birth defects. Many have been forced to abandon their homes, which have become virtually worthless on the resale market.

■ In both California and Kentucky, occupants of log homes that had been treated with the pesticide and wood preservative pentachlorophenol (PCP) have become ill, suffering such symptoms as numbness of the eyes, development of a blind spot, and nervous system impairment affecting their sleep, heartbeat, and appetite.

If you or a member of your family becomes ill and suffers from the effects of tainted indoor air, where do you turn for compensation? If the value of your home and property are undermined by steps you have taken that were recommended for saving energy, who should help make up the loss?

Home products and procedures a risk

These are questions raised by the increasing body of evidence that suggests that many products and procedures we are encouraged to use in our home have become serious threats to our health. As a result of the trend toward more tightly built buildings since the energy crisis of the early seventies, the U.S. government has been strongly advocating the use of energy-saving building techniques and retrofit procedures for existing high-energy-consuming homes, schools, offices, and public buildings. Indeed, market conditions added strength to the government's position, with spiraling fuel prices that further motivated citizens to take strong steps to reduce their energy usage. Initial research on the effects of energy-saving tactics gave assurances that residents and users of the buildings would be guaranteed a safe and healthy environment. Only more recent studies have uncovered the deleterious effects of energy conservation on indoor air and on the health of the people within the structures. Even now, the research remains insufficient, and much of the population remains uninformed about the nature of the problem and how to recognize it and deal with it in their own homes.

Redress Through the Courts

A chain of legal responsibility

Because of the time-consuming nature of promulgating new federal legislation, and because the concept of residential and nonoccupational indoor air quality regulation does not yet have wide understanding and support, the question of who will pay for losses and expenses in matters of damages caused by contaminated indoor air has mostly been settled in court under civil law. The legal grounds most often invoked in these cases are breach of contract or warranty, negligence, and strict liability.

When the adverse health effects are caused by the characteristics or building materials of the house itself, the homeowner as plaintiff may choose to sue the seller, the builder, the architect, and the manufacturers of the offending materials for financial remuneration. According to law, the seller is legally obligated to fulfill all contractual obligations express and implied, including the implied warranty of habitability that promises that a residential structure is fit to be lived in—that is, not a source of illness or danger. The builder is legally bound to construct a house in a careful and workmanlike manner, free from defects that would make it uninhabitable. The architect, as a professional, must have developed design plans with care, knowledge, and expertise. The manufacturer of building materials is expected to produce these materials carefully and without endangering defects. If the courts decide that any of these entities have failed in their obligations, they may then be civilly liable to the homeowner as plaintiff in the case.

Breach of contract or warranty

■ In the case of Blagg v. Fred Hunt Co., Inc. (1981), buyers who purchased a home within nine months of construction discovered a strong formaldehyde odor emanating from the carpet and carpet pad that had been installed by the builder. The court judged that the builder was liable for economic losses incurred by the buyers in replacing the carpet and pad. This case was particularly important because it extended the implied warranty of habitability to the third buyers of a house. The statement read, "A builder-vendor's implied warranty of fitness of habitation extends to subsequent purchasers for a reasonable length of time where there is no substantial change in the condition of the building from the original sale."[1]

■ In Waggoner v. Midwestern Development, Inc. (1968), water seepage into a basement led to the decision that where the seller of a new house is also the builder, there is an implied warranty of reasonable workmanship and habitability that survives the delivery of the deed. This was the first case in which a home was viewed as a consumer product and therefore had all of the implied warranties of fitness for its intended use.[2]

■ As early as 1952, in the case of Bradley v. Brucker, a builder-seller was sued after entering into an agreement that the basement of a home being sold to a person with a diagnosed pulmonary ailment would remain dry except for minor condensation. When water appeared in the basement only four months after the sale, and the condition was seen to be permanent, the builder-seller was held liable for breach of covenant in the sales agreement.[3]

■ Other cases of breach of contract or warranty involved products intended for home use, as in the case of Shirley v. The Drackett Products Company (1971), where the manufacturer and distributor of "Vanish" toilet bowl cleaner was held liable for respiratory damage suffered by the plaintiff who had breathed irritant gases released from the product while using it. Both the trial and appellate courts ruled that the product was not safe for its intended use and that this constituted a breach of implied warranty.[4]

■ In Alfieri v. Cabot Corp. (1964), the use of cooking charcoal for heating a poorly ventilated cabin resulted in the death of two residents by carbon monoxide poisoning. The court ruled that because the label on the charcoal said it was safe for cooking, indoors or outdoors, it could reasonably be foreseen as being safe for use indoors, and that the manufacturer was in breach of implied warranty of safety.[5]

Formaldehyde-related lawsuits

■ Today, one of the most active litigation battlefronts involves the use of formaldehyde in urea-formaldehyde foam insulation and particleboard. Literally thousands of individual lawsuits alleging personal injury from the formaldehyde exposure have been filed in the U.S., along with several class-action suits representing hundreds of thousands of other residents of homes in which UF insulation was installed. Many of these claims combine breach of contract, negligence, and strict liability theories.

Where individual cases have come to court, settlements have ranged from a few thousand dollars to $300,000. The manufacturer's knowledge of the potential hazards of the product and the failure to warn homeowners have been key factors in most of these cases. Both the plaintiff's physical injury and the loss of market value on the residence have been considered in arriving at compensatory and punitive damages. But unfortunately for the plaintiff, most of the companies producing the insulation have gone out of business since 1981, and many of the remaining ones are in bankruptcy, with insufficient insurance to cover the claims.

What to do

If you believe that you or a member of your family has suffered the effects of formaldehyde sensitization or has health symptoms relating to any product used in your home or work place, it is most advisable to talk with the Consumer Product Safety Commission. The CPSC has had experience in handling thousands of such complaints, and it can put you in touch with proper health officials, give you some suggestions on how to alleviate the problem, and even advise you on the basic legal issues. Some local and state health departments conduct tests for formaldehyde, radon gas, and other common pollutants, at little or no expense to residents. In addition, private companies specializing in pollution testing can be found in the Yellow Pages under "Laboratories." Discovering and documenting the level of the responsible pollutant (or pollutants) is the logical first step in treating the problem and seeking any financial redress.

Negligence and strict liability

A lawsuit based on negligence requires the plaintiff to prove that the defendant failed to take reasonable care in the manufacture and sales of its product. Because this can be difficult if not virtually impossible for the average citizen who is not privy to the details of manufacturing, many states—including New York and California—have adopted a "strict liability theory" of product liability. This allows a plaintiff to recover damages from a manufacturer or supplier of a product if that product is found to be defective or unreasonably dangerous, thereby causing injury. Many of the cases tried under this theory have involved carbon monoxide poisoning or illness caused by formaldehyde or asbestos.

■ In Heritage v. Pioneer Brokerage and Sales, Inc., and Moduline Industries, Inc. (1979), an Alaskan couple recovered damages from the retailer and manufacturer of their mobile home after suffering painful and disabling illness and lung damage caused by formaldehyde fumes in the mobile home. Although the original trial court had ruled in favor of the defendants, the Supreme Court ordered a new trial in which the jury was instructed to weigh the usefulness of the product (in this case, formaldehyde) against the risk of injury inherent in its use. In the end, the court ruled that the defendants had placed a defective and dangerous product into commerce.[6]

A variety of cases

■ In Harig v. Johns-Manville Products Corporation (1979), the court found the asbestos manufacturer strictly liable for the plaintiff's cancer, which occurred more than 20 years after exposure. Although there is normally a three-year statute of limitations in product liability cases, the court ruled that it did not apply here until after diagnosis of the disease, which in this case occurred 21 years after the last exposure. This decision is significant to the future of indoor air quality litigation because of the delayed nature of many of the effects of exposure to such contaminants as radon, asbestos, and the components of cigarette smoke.[7]

■ Cases involving accidental carbon monoxide poisoning include Wallinger v. Martin Stamping and Stove Company (1968), where although the victim installed the stove himself, the stove manufacturer was held liable for not specifying proper stack height.[8] In a similar case, Dover Corp. and J.R. Preis, doing business as Coastal Bend Sales, v. Perez, the manufacturer of the heater was found guilty of placing an unreasonably dangerous product in commerce. The company had failed to mechanically prevent placement of a 35,000 BTU heater in a 25,000 BTU heater case. In this instance, the owner of an apartment building mismatched the components, causing a carbon monoxide poisoning death.[9]

Federal Government Regulations

Although the courts have taken the brunt of the responsibility for the issue of indoor air quality, the government has not been completely unresponsive to the problem. To date, most of the research being conducted into indoor air pollution has been sponsored by federal and state governments.

Of course, precedents do exist for the government's taking action in regulating air quality to protect the public health and welfare. A decade of public outcry led to the passage of the Clean Air Act (or, more correctly, the Clean Air Amendments of 1970), which gave the Environmental Protection Agency (EPA) authority to impose limits on certain "criteria pollutants" found in ambient air, which the EPA has chosen to define as "air external to buildings" (see Table A for these standards). In addition, the U.S. National Institute for Occupational Safety and Health (NIOSH) regulates air quality in the work place (see Table B for selected work place standards).

Limited federal regulation

Still, despite cogent evidence that indoor air quality has a powerful impact on the public health, there are only two federal regulations governing indoor air quality in residential or commercial buildings, and these are conditional. One limits ozone emissions in houses, apartments, hospitals, and offices to 0.05 ppm by volume. The other concerns radiation levels, but only in homes built over former uranium processing or phosphate mining sites. Recent tests and surveys indicating an extensive radon hazard in the North-eastern states (especially Pennsylvania, New Jersey, and New York) have prompted the EPA to call 4 pCi/l the threshold of concern for indoor exposure to the gas, but no regulatory action is expected to be taken at present. Based on new data and improved methods of radon mitigation, this figure may soon be lowered, according to the agency.

Table A: National Ambient Air Quality Standards

	Long term		Short term	
Pollutant	Concentration (Micrograms/m^3)	Time	Concentration (Micrograms/m^3)	Time
Sulfur dioxide (SO_2)	80	1 year	365[a]	24 hours
Particulates	75[b]	1 year	260[a]	24 hours
Carbon monoxide	--	--	40,000[a] (or 35 ppm)	1 hour
Nitrogen dioxide (NO_2)	100 (or 0.05 ppm)	1 year	--	--
Ozone (O_3)	--	--	235[a] (or 0.12 ppm)	1 hour
Lead	1.5	3 months[c]	--	--

[a]Not to be exceeded more than once per year.
[b]Geometric mean.
[c]A calendar quarter.

Source: U.S. Environmental Protection Agency

Political pressures and the complexity of adding and enforcing new regulations discourage the EPA, or any other government agency, from moving to protect indoor air quality. Nevertheless, a number of existing federal laws—including the Clean Air Act, the Toxic Substances Control Act, and the Consumer Product Safety

Act—could be extended to regulate indoor air pollution. State and local laws can also be applied to this problem.

The Clean Air Act

As previously noted, this act allowed the EPA to set up the National Ambient Air Quality Standards (listed in Table A), which specify maximum allowable air concentrations for criteria pollutants, leaving the individual states with the responsibility of meeting these limits. If the EPA decided to apply the standards to indoor air, the states could choose to modify their building codes, impose emission standards on certain products, or restrict the sale and use of other products.

However, there are problems inherent in this approach. First, new standards would have to be formulated for the additional indoor pollutants of concern. This is a complex process involving expensive research, regulatory, and legal activities that could take several years. Second, if EPA simply extended its definition of "ambient air" to include the indoor environment, many air quality control regions—designated by EPA and each including several counties or even parts of several states—might suddenly find themselves not in compliance. They would be liable to stringent sanctions until reasonable methods could be developed for meeting the standards, both indoors and outside. One possible solution to this last problem could be for EPA to call indoor air a separate region in itself, giving the states time to develop a plan to achieve compliance.[10]

The Clean Air Act also authorizes the EPA to establish National Emission Standards for Hazardous Air Pollutants (NESHAPS), a rapid way of controlling substances that have been found to be especially dangerous. Under this authority, the EPA banned sprayed-on asbestos insulation and troweled-on decorative materials that contain asbestos. Currently pending legislation calls for the elimination of asbestos from roofing felt, flooring felt, linoleum and vinyl tile, and impregnated concrete pipe, and for a 10-year phase-down on all commercial use of asbestos, including importation, manufacturing, and sale of asbestos products. Other substances controlled under NESHAPS include beryllium, mercury, and vinyl chloride. If the EPA so desired, NESHAPS could also be set for indoor pollutants such as formaldehyde, which fall into the NESHAPS definition of substances that "may be reasonably anticipated to result in an increase in mortality or an increase in serious irreversible, or incapacitating reversible, illness."

The problem here is that not all indoor pollutants are immediately hazardous enough to meet this strict definition. Also, once labeled hazardous under this law, it would be difficult for EPA to explain why all emissions of the pollutant should not be banned—a situation that would often be economically and practically unfeasible.[11]

The Toxic Substances Control Act

The Toxic Substances Control Act (TSCA) of 1976 gives the EPA the right to regulate chemical substances that "present an unreasonable risk of injury to health or the environment." Under this law, the EPA can prohibit or limit the manufacture, processing, or distribution of a chemical entirely or for specified uses; require manufacturers to give notice of potential health risks, provide instructions for the substance's use, or replace or repurchase the substance; and order manufacturers to improve quality control standards. The EPA must invoke the least burdensome of these requirements to protect the public adequately from health risks.

Formaldehyde and asbestos are two indoor pollutants that could be regulated under this act. Exposure to formaldehyde, for example, could be limited by banning or restricting the use of formaldehyde insulation, particleboard, carpets, fabric, or furniture containing formaldehyde either completely or under certain conditions, such as in homes with low ventilation rates. The Environmental Protection Agency could also order manufacturers of UFFI to remove and replace their product in such homes, or to improve quality control so that formaldehyde emissions are substantially eliminated.

Applying the law

The major difficulty in applying the TSCA to indoor air is in getting the EPA to act, because of the high cost of the necessary research and listing procedures. According to the agency, it can cost up to a quarter of a million dollars to develop the testing procedure for a single substance, and this sum does not include the cost of developing regulations based on the test results. Under TSCA, the EPA considers its main obligation to inventory the existing chemical substances in commerce (more than 55,000 of them), to enforce manufacturers' recordkeeping, testing, and reporting requirements, and to obtain premanufacture notice of proposed new chemicals. Pollutants that have had action taken under TSCA include chlorofluorocarbons (spray propellents), polychlorinated biphenyls (PCBs), chlorinated benzenes, chloromethane, and phthalate esters.[12]

The Consumer Product Safety Act

The Consumer Product Safety Act (CPSA) of 1974 authorized the Consumer Product Safety Commission (CPSC) to regulate consumer products, which the commission defines as products "for use in and around" a residence. That could include stoves, portable heaters, carpeting, appliances, and insulation, and possibly construction materials, which are responsible for emission of pollutants.

The CPSC can set safety standards for a consumer product to prevent or reduce chances of injury from its use but must rely on voluntary standards from manufacturers whenever those standards would adequately provide the same service. The CPSC can also ban a product it feels would present an unreasonable risk of injury, if no standard would be seen to offer adequate protection to the product's users.

Thus, emissions from gas stoves, for example, could be regulated, as could formaldehyde emissions from home furnishings or construction materials. Kerosene heaters could be banned or made to carry warnings about the dangers of their use. Radon emissions from bricks, concrete blocks, and building stone could also be limited under this law. In 1970, the CPSC did restrict radiation emissions from color televisions to less than 0.5 milliroentgens per hour (0.5mR/hr) and in 1971 limited allowable radiation leakage from microwave ovens to less than 5 milliwatts per square centimeter (5 mW/cm^2). Both dosages are generally considered safe.

Problems with CPSC

Again, there are problems with regard to covering indoor air pollution under this act. First, it is arguable whether houses can be considered "consumer products" and whether building materials within houses fall under that heading. Some courts have ruled that they are, but the question is far from adequately answered. Second, ventilation standards, which are often the key to indoor air quality, are not within the scope of this law. Third, the CPSC can move only against products that can cause imminent illness, but not against those that may cause adverse health effects in the long term. Other problems include the CPSC's inability to regulate consumer use of products, and certain limitations with regard to its joint jurisdiction with the Clean Air Act, the Occupational Safety and Health Act, and the Atomic Energy Act.[13]

In one major move against an indoor air pollutant—urea-formaldehyde foam insulation—a 1982 CPSC ban on the product was subsequently overturned by the U.S. Court of Appeals in New Orleans on grounds of "insufficient evidence" of its threat to the public's health. The insulation was permitted back on the market. In 1984, however, the EPA announced that it would begin studying the relationship between formaldehyde and cancer under its mandate in the Toxic Substances Control Act. Its assessment of whether formaldehyde presents an "unreasonable risk" to human health is focused on people living in homes built largely of formaldehyde-containing materials, and textile and clothing workers who deal extensively with textiles treated with formaldehyde resins.[14]

State and Local Government Legislation

Responsive legislation

State and local governments generally have been more responsive to the need for maintaining healthy standards of indoor air. In 1980, Minnesota became the first state to pass legislation regulating formaldehyde levels in residences, followed in 1981 by Wisconsin. Other states, including New York, New Jersey, Massachusetts, West Virginia, South Dakota, and Arizona have also moved to control indoor air quality. California has implemented the most comprehensive indoor air quality program to date, mandating, for example, that the tightest of new residences must achieve a minimum of 0.7 air changes per hour, and that workplaces must be monitored to ensure the ASHRAE Standard 62–1981, regarding ventilation, is met or exceeded in building operation.

Under the Asbestos Hazard Emergency Response Act, states have implemented programs for building inspection and management, asbestos abatement and accreditation programs and enforcement of asbestos standards. New Jersey's program goes beyond federal standards, requiring that only trained and approved professionals undertake removal actions and that airborne asbestos levels remain within limits during removal activities.

Many states have also undertaken programs to study and mitigate radon problems in buildings. Leaders in this area include Florida, New Jersey, New York and Pennsylvania, precisely those states that first encountered dramatically elevated radon levels in many of their homes.

In 1983, Oregon became the first state to enact emission-control legislation for wood stoves. According to the law, by July 1986, all new wood stoves sold in Oregon had to achieve a reduction in particulate emission of 50 percent, compared with stoves sold in the 1970s and early 1980s. By July 1988, the reduction had to be 75 percent in order for new stoves to be sold in the state. These strict standards were prompted by a study during the winter of 1980–1981, which attributed between 66 percent and 84 percent of inhalable particles in ambient air to wood stove use. More than half the homes in Oregon are heated with wood to some extent.

In 1987 the federal government emulated Oregon, and under the mandate of the Clean Air Act developed nationwide standards for wood stove

Table B: Selected Work Place Safety and Health Standards

Pollutant	Concentration[a] (ppm)	mg/m³
Ammonia	50	35
Carbon dioxide (CO_2)	5,000	9,000
Cresol	50	55
Formaldehyde	2	3
Nitric oxide (NO)	25	30
Nitrogen dioxide (NO_2)	5	9
Ozone (O_3)	0.1	0.2
Propane	1,000	1,800
Sulfur dioxide (SO_2)	5	13
Trichloromethane	50	240
Respirable dust	--	5
Asbestos	Fewer than two fibers	>5 micrograms/cc

[a]Applicable to an 8-hour time-weighted average, except for SO_2,which is a ceiling value.

Source: National Institute of Occupational Safety and Health, as cited in Meyer, Beat: *Indoor Air Quality*

emissions. Wood stoves can now be purchased that are certified to meet these standards.[15]

The states can also help regulate indoor air quality through their power to issue building codes. Such codes have the potential to specify low-emission materials, recommend special aging procedures, or even boost required ventilation rates and lighting requirements. They could also dictate site preparation procedures in cases of development on landfill or mine tailings areas, or in radon hazard areas, and other installation techniques.

Smoking vs. the Law

One area in which regulatory gains are being made, despite court challenges, is that of smoking in public or work places. One of the most stringent of these laws was adopted by Nassau County (New York) in 1986. It requires Nassau County restaurant owners to install a specific ventilation and air-cleaning system in their restaurants or to set aside half their seating for nonsmokers. Alternatively, the restauranteurs could install a lesser ventilation system and set aside one fourth of their seating for nonsmokers. The law also requires an employer to designate a "separate portion or portions of the work area, employees' lounge, and cafeteria for smoking," leaving the rest of the work area smoke-free. Any employee could assert his or her rights to a smoke-free work space by complaining to the Board of Health about infringement by a smoker. The board would send the offender a letter warning that he or she was breaking the law and could be subject to a hearing before the board and a $250 fine.

The Nassau County regulation was said to have been strongly influenced by neighboring Suffolk County's antismoking ordinance, which provides that nonsmokers can declare their immediate work areas to be no smoking zones and mandates that companies with more than 75 employees separate smokers and nonsmokers. It also requires restaurant owners with 50 or more seats to reserve 20 percent of them for nonsmokers.[16]

In deference to the emerging realization of the importance of clean indoor air, several states and localities including New Hampshire, Michigan, Maine, Minnesota, Los Angeles and Montgomery Co., Maryland, have actually enacted legislation revising and strengthening non-smoking laws that were in effect before 1987. In many cases the new legislation goes as far as banning the smoking of tobacco in such places as childcare and health facilities, and strengthening the provisions that apply to private workplaces.

The single greatest source of respirable particulates

Regulating peoples' right to smoke has always been an extremely emotional issue, as it appears to dictate against personal freedom. But tobacco smoke has been found to be the single greatest source of respirable particulates in the home, and a significant source of carbon monoxide, benzo(a)pyrene, and aromatic hydrocarbons, and the Surgeon General of the United States, in 1974, amended its original assessment of a decade earlier to include the warning that passive smoking, too, may lead to chronic disease, including cancer.

This is not the first time smoking has been regulated. In the late nineteenth century, some 14 states prohibited the recreational use of tobacco, calling it a fire hazard and a "noxious and unhealthy habit, wholly harmful and without any virtue whatsoever." By the early twentieth century, however, most antismoking ordinances were overturned.

Even after the 1964 Surgeon General's Report linking smoking with cancer, public opinion held that smokers had a right to run their own risks. In 1973 the Civil Aeronautics Board (CAB) passed a rule requiring airplanes to reserve separate seating for nonsmokers, but almost no one on the planes was willing to enforce it. Then in 1976 Allegheny Airlines agreed to offer passengers a choice of smoking or nonsmoking seating and to disembark any passenger who persisted in smoking in the wrong section, and the other airlines subsequently followed suit.

In April 1988, a federal law went into effect banning smoking on flights of less than two hours, and carrying stiff fines of up to $1,000 for smoking on such flights and up to $2,000 for tampering with lavatory smoke detectors. This ban applies to some 80 percent of all U.S. flights, or an estimated 13,6000 flights per day.

In 1976 the Appellate Court of New Jersey ruled in the landmark case of Shimp v. Bell Telephone that employees have the right to a smoke-free work place. The decision was based on compelling evidence that sidestream smoke is indeed toxic to a significant degree, and that employers have the obligation of providing a reasonably safe work place.[17]

Change in public attitude

The 1976 cases marked a turning point in public attitudes toward smoking. By 1980 seven states —California, Connecticut, Maryland, Montana, Nebraska, Oregon, and Rhode Island—had limited smoking in public spaces, and in that same year the U.S. Department of Defense (DOD) banned smoking in its public areas and established nonsmoking eating areas. Today, 45 states (including the District of Columbia) and more than 460 localities have restrictions against smoking in such places as public transit vehicles, elevators, government buildings, retail and grocery stores, health care facilities, schools, museums, concert halls, sports arenas, and at public meetings.

Smoking in the home remains unregulated and almost unapproachable from a legal perspective, although tobacco companies have recently come under attack from the smokers themselves or their survivors, who are charging the producers with liability for chronic diseases and physical injury or even wrongful death connected with years of tobacco use. To date, the courts have ruled mainly in favor of the corporations, such as in the 1985 dismissal of a case against R.J. Reynolds Tobacco Company, in which a Tennessee man claimed that a lifetime of smoking was responsible for circulation problems that led to the amputation of his left leg in 1983, or the 1985 ruling that R.J. Reynolds was not liable for the death of a California man who smoked three packs a day for more than 50 years.[18]

Recently, however, the tide has taken something of a turn. In June 1988 a federal jury found the Liggett Group, a cigarette manufacturer, liable in the lung cancer death of a New Jersey woman, Rose Cipollone. The jury concluded that the company had failed to warn of the health risks of smoking and had, in fact, used slogans such as "Just what the doctor ordered" in advertising campaigns before the 1966 mandate that required cigarette packs to carry a health warning. Such advertising, the jury found, misled the public by suggesting that smoking was safe. The jury then awarded $400,000 in damages to Antonio Cipollone, husband of the deceased.

Despite the fact that this case marked the first time since 1954 in which a tobacco company had lost a claim or been ordered to pay even one cent in damages, its landmark status was not considered a major setback by the tobacco industry since the court also found Ms. Cipollone herself 80 percent responsible for her own death from cancer. Cipollone had smoked 1½ packs of cigarettes per day for some 40 years. Indeed, the jury found Liggett and two other cigarette manufacturers (Lorillard and Phillip Morris) not guilty of the charge that they had fraudulently misrepresented the risks of smoking and had conspired to misrepresent the facts.

The significance of this case was further eroded early in 1990 when a federal Court of Appeals for the Third Circuit (in Philadelphia) struck down the award, ruling that the Cipollone family had not proven that Ms. Cipollone had seen any of Liggett's ads.

In another sign of hope for tobacco liability plaintiffs, a New Jersey Supreme Court ruled in July 1990 that warning labels on cigarette packs and advertisements do not protect manufacturers against product liability lawsuits filed by smokers or their relatives. Going against five federal appeals courts decisions that tobacco makers were immune to smoker-death lawsuits filed after the 1966 mandate for such warnings, the New Jersey court ruled that Congress did not intend to "immunize" cigarette manufacturers against suits simply by imposing the warning labels.

The decision stemmed from a suit filed in 1982 by Claire Dewey against Brown & Williamson Tobacco Corp., American Brands Inc. and R.J. Reynolds Tobacco Co.. Her husband, Wilfred Dewey, died in 1981 of lung cancer at age 49 after 40 years of smoking. The court ruling allows Dewey to proceed with the product liability case. If the cigarette makers decide to appeal this decision at the U.S. Supreme Court level they risk losing because that court has a track record against federal preemption. If they do not appeal, they risk future law suits stemming from deaths of smokers who purchased cigarettes after 1966.

Need for public action

Clearly, government intervention in the matter of indoor air quality will be no easy task, especially at the present time when the federal government appears to be abdicating its role in matters of social and environmental concern. Once again, it is up to the public to become educated on this issue and to provide the political impetus to move the government to action.

6 Close-up on the Pollutants: The Pollution Compendium

An old joke tells of a woman at the zoo who reads a sign: "This way to the Egress." She follows the sign, eagerly looking for the strange and exotic egress, only to find herself outside the gates of the zoo.

We laugh at her naive lack of verbal comprehension, acutely aware of our own shortcomings in this and similar areas. But in the case of indoor air quality, we all need to become as well informed as possible in order to safeguard our health. We do not need to be alarmed, but only aware of the characteristics and potential adverse health effects of the many substances that share our dwelling and work space, so that we may make wise decisions about actions to take to avoid danger—imminent and future.

Increase awareness

This chapter is designed to provide in-depth explanations of the properties of many of the common pollutants, the range of concentrations in which they have been found indoors, and proven and suspected adverse health effects—a pollution compendium to add to your understanding of the issue of indoor air quality. The substances are listed alphabetically, except in the case of Organic Chemicals, which are all grouped under that category.

Aerosol Propellents

Application of all manner of household products,

from pesticides to oven cleaners to personal hygiene products to dusting sprays, has been made simpler by the use of aerosol spray cans, which have achieved increasing popularity since the 1950s. Until recently, these aerosols were propelled by chlorofluorocarbons (Freon 11 and Freon 12), which have been judged to have a deleterious effect on the ozone layer of our atmosphere. They are no longer in use.

Hydrocarbons

Today, spray cans are propelled mainly by volatile hydrocarbons (propane and butane), nitrous oxide, methylene chloride, and other gases. When aerosol products are used, significant amounts of propellent are released into the air, as noted earlier. More important, the ingredients in the product itself, which are often irritating or even toxic, are released in a fine mist of which a substantial portion hangs suspended in the breathing area.

Adverse health effects

Nitrous oxide and methylene chloride have an anesthetic effect on the central nervous system. When methylene chloride is inhaled it can be converted to carbon monoxide, which produces its own toxic effect on the body. In addition, the Food and Drug Administration (FDA) believes it has sufficient evidence of the potential carcinogenicity of methylene chloride to ban it from such consumer products as hair spray. Other propellents are currently under study for suspected ill effects. In addition, some propellents are highly flammable and can easily ignite when used extensively in poorly ventilated conditions.

Allergens

Allergens are another diverse group of pollutants, bound together by the immune response they produce in the body. Common reactions to allergens include stuffy nose, sneezing, coughing, wheezing, asthma, itching, rash, even malaise and headache. Within the home, the most common carrier of allergens is house dust. As opposed to highway dust, which is high in metals such as lead, copper, zinc, and iron, house dust consists mainly of animal dander, human skin scales, insect excreta, food remnants, bacteria, radioactive dust, and sometimes molds and pollen. The excreta of a particular microscopic insect that lives off human sebum keratin—the dust mite—often produces allergic rhinitis and asthma. These dust mites commonly abide in the bedroom, where they live most productively between the sheets. Making and changing the bed helps disperse them throughout the room and the rest of the house.

Animal dander

In homes with pets, animal dander, particularly from cats, dogs, rabbits, and guinea pigs, is another important cause of allergic rhinitis and asthma. Pollens, which are seasonal and most often come from outdoor plants and trees, are particularly universal allergens, especially the pollens of grass and ragweed. Fungal spores and molds, which are prevalent at high humidities, also trigger allergic responses in sensitive people.

Adverse health effects

Allergic reactions in the respiratory tract often follow inhalation of the above-mentioned al-

Allergic responses

lergens. The size and penetration of the particulate in question generally indicate the nature of the reaction. Allergic rhinitis is a local inflammatory response involving primarily the nose and is most often triggered by the larger particles, such as pollens (15 to 35 micrometers). Allergic asthma partially narrows the tracheobronchial airways and is often associated with particles in the 5- to 10-micrometer range, such as molds and organic dusts (including animal and insect excreta), although pollen grains have also been implicated. Smaller particles (from 2 to 4 micrometers), which penetrate to the alveoli, may trigger seriously threatening allergic responses, such as hypersensitivity pneumonitis (involving inflammation of the alveolar walls and bronchioles) or they may be eliminated from the lungs. Bacteria and amoebae found growing in humidifiers and ventilation systems, and minute particles of the feces of birds such as parakeets and pigeons, have been known to elicit this response, which is sometimes life-threatening.

Asbestos

Used extensively in construction

Asbestos is the collective term used to describe several silicate mineral fibers that tend to be flexible, durable, incombustible, and good thermal and electrical insulators. Under a microscope they generally look needlelike—long, thin, and sharp—hence their ability to embed themselves deeply in human tissue. Those that are 0.1 to 5 microns in size are readily inhaled and lodge in the lung and lymph ganglions, where they may remain permanently. Because of its fireproof nature and cheap availability, asbestos was used extensively in all types of construction until about 1960, when its long-range health effects—on workers and building users—began to be known. Asbestos has also been used in home appliances such as toasters and hair dryers, stoves and furnaces, and in spackling compound and building facade material such as asbestos shingles (see chapter 2). Most of the asbestos originally used in construction remains in place, except where local, state, or the federal government has ordered it removed, as in some schools. Once released from its binding material (such as by erosion, vibration, renovation, or vigorous cleaning) the fibers can remain airborne for long periods.

Concentrations See Table A for concentration levels.

Adverse health effects Asbestos fibers that lodge in the lungs are responsible for asbestosis, a nonmalignant respiratory disease characterized by dry cough, labored breathing, shortness of breath, pleural thickening of the lung tissue, mesotheliomas (tumorous growths), and fibrosis. Malignant lung cancer can subsequently develop 12 or more years after the fibers enter the lungs, even with relatively low exposures (as in the families of asbestos workers exposed to contaminated clothing). Heavy exposure, especially among smokers, increases the risk of mortality. There is no known cure for asbestosis. According to a 1979 EPA survey, anyone exposed to more than five fibers per milliliter of air for any length of time is in a high risk situation with respect to asbestosis.

Table A: Airborne Asbestos Concentrations in 11 Locations (determined by electron microscopy count)

(OSHA standards allow 2.0 fibers/ml averaged over an 8-hour day)

Location and activity	Mean (ng/m^3)	Range (ng/m^3)
Urban outdoors		
48 U.S. cities	<10	—
New York City	17	2-65
Indoors[a]		
Connecticut office building (exposed sprayed ceiling)	79	40-110
Connecticut grammar school (exposed ceiling, with custodial activity)	643	186-1,100
New York City schools	99	9-135
Massachusetts schools	151	38-260
10 New York, New Jersey, and Massachusetts schools with damaged asbestos surfaces	217	9-1,950
New Jersey apartment building (heavy housekeeping)	296	—
New York City office buildings (asbestos in ventilation systems; routine activity)	2.5-200	0-800

1 nanogram (ng) = one billionth of a gram

From National Research Council, *Indoor Pollutants*, and Wadden, Richard A., and Scheff, Peter A.: *Indoor Air Pollution*

Carbon Dioxide

Carbon dioxide (CO_2), an odorless, colorless gas, is a normal by-product of human, plant, and animal metabolism and is therefore rarely considered a pollutant. Because it is exhaled from the lungs at a rate of about 200 milliliters per minute (200 ml/min.) by the human body, the carbon dioxide level of occupied rooms is always higher than in unoccupied rooms (all other sources being equal). Because the gas is chemically inert, it does not react with building materials and remains in the air until it is diluted or removed through ventilation with clean air.

Concentrations

Typical outdoor levels range from 100 to 500 ppm, while typical indoor concentrations are from 50–1200 ppm, depending largely on human occupancy.

Adverse health effects

At high concentrations (above 1,000 ppm), carbon dioxide may cause headaches, poor judgment, shortness of breath, and drowsiness. At extremely high levels (where air contains more than 6 percent CO_2), unconsciousness will occur within minutes.[1] Although serious effects of carbon dioxide are rare in normal home, apartment, or office situations, CO_2 levels can become high enough to be toxic in such places as submarines (which recycle virtually all their air at times) and in farming and wine production areas where the gas settles and accumulates in basements, vaults, and wells.

Carbon Monoxide

One of most dangerous pollutants

Carbon monoxide (CO), a colorless, odorless, toxic gas formed by the incomplete combustion of fossil fuels, is the most prevalent of indoor pollutants, and Dr. John Spengler of Harvard University believes it remains one of the most dangerous. It forms in poorly tuned gas and oil furnaces, ranges and ovens, wood- and coal-burning stoves, and parking garages when automobiles are left running. Ice cleaning and finishing equipment in skating rinks are responsible for high levels of carbon monoxide in indoor arenas. Poorly ventilated kitchens, rooms over garages, buses, and automobiles with leaky exhaust systems can be sites of dangerously high levels of this gas.

Concentrations

Air levels of CO are greatly affected by use of unvented combustion appliances (stove, ovens, heaters, and by the presence of tobacco smoke). It is estimated that a gas range produces 1 to 3 mg of CO per hour of operation, and that it can increase CO levels by a factor of 15 over ambient (outdoor) concentrations.[2] Cigarettes emit 25 to 50 mg of CO in sidestream smoke, and in poorly ventilated rooms with smokers, levels can reach peaks of 40 to 80 ppm, resulting in symptoms of acute poisoning, according to a study by the U.S. Department of Health, Education and Welfare.[3] In many homes and office buildings, attached garages constitute the strongest source of CO. Heavy traffic can produce CO levels of up to 140 ppm (for short periods of time), and commuters are often exposed to levels of 50 ppm or more (see Table R).

Adverse health effects

Carbon monoxide interferes with the blood's ability to transfer oxygen throughout the body. It binds to hemoglobin in the blood more strongly than oxygen does, forming carboxyhemoglobin (COHb), which cannot carry oxygen to the cells. The symptoms of carbon monoxide poisoning include headache, impaired vision and judg-

CO poisoning

ment, and irregular heartbeat. Symptoms of acute poisoning can result from concentrations above 20 ppm. In nonsmokers, the normal level of COHb in the blood is about 0.4 to 0.7 percent although ambient air pollution or exposure to sidestream smoke can raise that level to two to three percent. Smokers typically have COHb saturations of seven to eight percent. In patients with cardiovascular disease, even small (four to five percent) acute increases in carboxyhemoglobin levels could exacerbate their symptoms. Adverse physiological effects begin at a carboxyhemoglobin level of about 2.5 percent, which can be reached in 90 minutes in air with 50 parts per million (ppm) CO, or in 10 hours at 15 ppm. By the time blood levels reach 10 percent there is significant reduction in visual perception, manual dexterity, and ability to learn.[4] Exposure to over 500 ppm for more than one hour can lead to approximately 20 percent carboxyhemoglobin saturation. Exposure to 1,500 ppm for one hour is usually fatal.[5]

Formaldehyde

Although, technically speaking, formaldehyde is an organic compound (that is, it consists largely of carbon, along with hydrogen and oxygen) its extensive presence in the indoor environment and the controversial nature of its health effects make it worthy of individual discussion. Aldehydes in general are a specific group of organic compounds that vaporize at room temperature. Formaldehyde, the simplest of the aldehydes, is colorless and has an irritating, pungent odor noticeable at less than 0.1 milligram (mg)/cubic meter. In a room, formaldehyde can be smelled at 1 part per million (ppm), but sensitive people can detect it at levels as low as 0.05 ppm.

Ubiquitous presence in home and office

With the widespread use of fiberboard, particleboard, plywood, and wall-to-wall carpeting the presence of formaldehyde has become ubiquitous in the American home and office. It is also found in the home in such sources as tobacco smoke, emissions from stoves and heating systems, textiles, and paper products. Formaldehyde (especially in the form of urea-formaldehyde foam insulation) has been the source of literally thousands of complaints to the Consumer Product Safety Commission.

Concentrations

For concentration levels, see Table B.

Adverse health effects

Formaldehyde is highly irritating to the eyes, skin, and mucous membranes. (See Table C.) Because it is water soluble it can easily enter the bloodstream. The presence of even low concentrations of formaldehyde in the air can cause burning and tearing eyes, headaches, drowsiness, nausea, vomiting, diarrhea, and irritation of the nose and throat. Concentrations of more than 5 ppm are likely to produce coughing, wheezing, and chest constriction. At concentrations of 50 to 100 ppm, pulmonary edema, pneumonitis, and death are likely to result. Formaldehyde has also been shown to trigger bronchial asthma attacks. Skin contact with for-

Table B: Selected Examples of Observed Formaldehyde Concentrations (The instantaneous ambient air quality limit is 0.12 mg/m^3 or 0.097 ppm.)

Sampling site	Concentration, ppm Range	Mean	Source
Two mobile homes in Pittsburgh, Pa.	0.1-0.8	0.36	(1)
Sample residence in Pittsburgh	0.5 (peak)	0.15	(1)
Mobile homes registering complaints in Washington	0-3.0	0.4	(1)
Mobile homes registering complaints in Minnesota	0.02-4.2	0.88	(1)
Public buildings and energy-efficient homes	0-0.21	—	(1)
28 residences with UFFI	0.02-0.13	0.07	
control group (non-UFFI)	0.03-0.07	0.05	(2)
78 structures:			
apartments	—	0.08	
homes (UFFI and non-UFFI)	0.03-0.20	0.05	
public buildings	—	0.04	(2)
3 residences:			
UFFI	0.11-0.16	—	
non-UFFI	0.06-0.08	—	
energy-efficient, non-UFFI	0.13-0.17	—	(2)
164 mobile homes (1984)	<0.02-0.78	0.15	(2)
65 mobile homes (registering complaints) (1979)	<0.01-3.68	0.47*	(2)
65 mobile homes (1979)	<0.10-0.80	0.16*	(2)

Sources:
(1) National Research Council: *Indoor Pollutants*. p.IV 31.
(2) Sterling, David A.: "Volatile Organic Compounds in Indoor Air: An Overview of Sources, Concentrations, and Health Effects," *Indoor Air and Human Health*, Gammage, R.B., and Kaye, S.V., eds. p. 389.

*median

maldehyde may cause irritation, contact dermatitis, and urticaria (hives).[6]

Repeated exposure to formaldehyde may cause an allergic sensitivity to the substance to develop. In this case even minute concentrations of formaldehyde can trigger an allergic response in the individual, long after the problem caused by initial exposure has been alleviated.

Table C: Reported Health Effects of Formaldehyde at Various Concentrations

Effects	Approximate Concentration (ppm)
None reported	0.0-0.05
Odor threshold	0.05-1
Neurophysiological effects[a]	0.05-1.5
Eye irritation	0.01-2.0[b]
Upper airway irritation	0.10-25
Lower airway and pulmonary effects	5.0-30
Pulmonary edema, inflammation, pneumonia	50-100
Death	100+

[a] Changes in electroencephalograms and in the response of dark adapted eyes to light.
[b] The response to 0.01 ppm occurred in the presence of other pollutants.

Source: National Research Council: *Formaldehydes and Other Aldehydes,* Washington, D.C. National Academy Press, 1981.

Cancer and formaldehyde

Controversy rages about the possible carcinogenic effects of formaldehyde. Research performed on rats and mice by the Chemical Industry Institute of Toxicology found that exposure to 15 ppm of formaldehyde, six hours a day, five days a week induced nasal cancers after 11 months. A lesser dosage produced structural changes in the nasal mucous membranes, which along with the cilia perform the important task of filtering and clearing foreign material from the nose. If this process is disturbed, other harmful material, normally kept from the airways, can enter the body. From this and other data, the EPA in 1987 issued a risk assessment of formaldehyde that formally classified it as a "probable human carcinogen." Additional studies are ongoing and new risk guidelines are expected to emanate from the agency as the data is analyzed.[7]

Infectious Agents (Aeropathogens)

Damp surfaces a breeding ground

Disease-producing agents such as viruses, fungi, and bacteria abound in the home and work space. Each human sneeze releases up to 40,000 droplets, each teeming with microbes, that travel at speeds of up to 40 meters per second. Each of the million human skin scales we shed daily carries an average of four bacteria. Damp walls, rugs, decaying wood, air conditioners, humidifiers, toilets, frost-free refrigerators are all breeding grounds for infectious agents.

The importance of airborne infectious agents was acknowledged in the last century in regard to the spread of pulmonary tuberculosis. It has since been discovered that just one tuberculosis bacillus in 50 m^3 of air is sufficient to cause the disease. Today, we are lulled into thinking that with increased sanitation and polite personal habits we are less vulnerable to attack from airborne microbes, but we are not. Buildings with low ventilation rates (highly recycled air) have become sites of recent epidemics of such dangerous illnesses as Legionnaire's disease (caused by a bacterium), measles, Q-fever (caused by a rickettsia), influenza (caused by a virus), and a host of other, unspecified diseases. This is the source of the term "sick-building syndrome," which is drawing increased attention from government health agencies and the media.

Nitrogen Oxides

Nitrogen oxide (NO) and nitrogen dioxide (NO_2) are combustion by-products produced by the burning of natural gas or oil in oxygen-rich environments such as kitchen stoves and ovens, furnaces, unvented gas and kerosene heaters, automobile engines, coal heaters, etc. It is also produced in cigarette smoke (see appropriate tables in "A Walk Through the House" for emission rates). While nitrogen oxides were a key component of urban smog before the use of pollution control devices in gas-powered vehicles, their primary indoor sources are gas stoves.

Concentrations For concentration levels see Table D.

Adverse health effects Nitrogen oxides are respiratory irritants at low concentrations and can cause lung damage and even death at high concentrations. Concentrations of NO_2 above 282 mg/m^3 (equal to 150 ppm), can cause acute pulmonary edema and death, levels of 94–282 mg/m^3 (50–150 ppm) can produce chronic lung disease (bronchiolitis), and exposures above 47 mg/m^3 (25 ppm) can cause bronchitis. Odor can be perceived as low as 0.23 mg/m^3 (0.12 ppm). In animal studies, nitrogen dioxide exposure significantly increased mortality of animals challenged with bacterial aerosols, and researchers also found a possible connection between NO_2 exposure and a propensity toward emphysema.[8]

Table D: Observed Levels of Nitrogen Dioxide (Work place limit is 900 micrograms/ cubic meter or 5 ppm; annual ambient air quality standard is 100 micrograms/ cubic meter or 0.05 ppm.)

Location	Concentration (micrograms/cubic meter)	Source
Kitchens with gas stoves	Up to 1,000 with one top burner operating for less than 30 minutes; Up to 1,700 with oven on for 20 minutes	(1)
	200 one meter from the stove, with only pilot on; 500-1,200 with stove in use	(2)
Bedrooms of house with forced air gas-fired heating system	Up to 1,200 for 8 hours	(1)
24-Hour Mean Concentrations (in parts per billion)		
37 electric kitchens in Chicago	40	(2)
121 gas kitchens in Chicago	74	(2)
69 gas kitchens in Columbus, OH	54	(2)
50 electric kitchens in Columbus	18	(2)
Outdoor levels (Columbus)	29	(2)

Sources:
(1) National Research Council: *Indoor Pollutants*
(2) Meyer, Beat: *Indoor Air Quality* p. 174.

One of the most significant effects of nitrogen oxides may be the creation of slight, almost imperceptible respiratory changes in infants and young children, indicated by increased respiratory infections. Dr. John Spengler of Harvard University believes the significance of this may lie not so much in the health of the child but in the aging process of the future adult. In England, R.J. Melia, C. du V. Florey, and S. Chinn did several studies of the effects on children of cooking with gas stoves. Increased incidence of bronchitis, cough, deep chest colds, wheeze, and asthma were seen in three separate studies of children between the ages of five and eleven.[9]

Respiratory irritants

Another researcher, Frank Speizer of Harvard, found that among some 8,000 children from six to ten years of age, those from households that cooked with gas experienced a more extensive history of respiratory ailments before the age of two. He also found that children from these homes had decreased lung functioning when tested for volume of air exhaled. Subsequent studies by the EPA and Johns Hopkins University echoed these results. Measurements taken over 24 hours in representative homes have found that NO_2 levels are two to three times higher in homes with gas stoves than in homes with electric stoves.[10] Other studies have found no correlation between gas cooking and lung impairment or increased respiratory illness in children, although in general these studies have tested much smaller populations of children. Studies of homemakers in houses with gas stoves have not found evidence of increased respiratory disease in adults.

Organic Chemicals

Organic chemicals are those compounds having carbon as a major constituent of their chemical makeup. This group includes the plentiful halocarbons and hydrocarbons found in petroleum distillates. Volatile organic chemicals (VOCs) are those that vaporize readily at relatively low temperatures (such as might be encountered in the home). This makes the category of "organic chemicals" a large one indeed. More than 250 volatile organic compounds have been found in quantities of more than 1 part per billion (ppb) in indoor air of urban residences and offices, many of which are potentially toxic or carcinogenic in sufficient quantities.

VOCs and PVCs

In a recent study of 40 Tennessee homes, funded by the Consumer Product Safety Commission (CPSC), between 20 and 150 such chemicals were identified in the test homes, as opposed to 10 or fewer outside the homes, and the levels of those that appeared in both places were generally 10 times higher indoors than outdoors. The high levels of organic compounds found in the homes were attributed to aerosols, cleaners, polishes, plastics, paints, varnishes, furnishings, heaters, pressed wood products, and pesticides, among other items.

For example, automobile upholstery can release more than 147 different organic compounds within the car's tightly enclosed space, many of them toxic, including the highly toxic vinyl chloride monomer, found in the form of polyvinyl chloride (PVC), a highly durable plastic which outgasses this substance readily during the first three months of new car use. It is this substance that is generally considered responsible for the typical "new car smell." The EPA has set a target level of zero for the vinyl chloride monomer because of its toxic effect on workers in the PVC industry.

Table E: Levels of Selected Organic Chemicals

Chemical	Winter mean level (micrograms/cubic meter)	Summer mean level (micrograms/cubic meter)
Toluene	27.2	61.7
Ethyl benzene	4.4	10.5
Xylene	16.6	44.2
Nonane	10.6	6.4
Cumene	2.2	1.3
Benzaldehyde	22.3	51.7
Mesitylene	1.4	6.7
Decane	12.4	8.9
Limonene	10.3	20.8
Undecane	8.2	12.0
Napthalene	9.4	17.5
Dodecane	4.2	13.7
2-Methylnaphthalene	2.5	2.6
Tridecane	2.3	16.5
Tetradecane	3.2	7.5
Pentadecane	2.8	1.9
Hexadecane	3.7	3.8
Benzene	50.0	100.0

One microgram = one millionth of a gram

From U.S. Consumer Products Safety Commission: "Status Report on Indoor Air Quality Monitoring Study in 40 Homes"

Another group of organic chemicals, the "highly volatile organics" (such as halogenated hydrocarbons, aldehydes, methylene chloride, trichloroethane, and perchloroethylene), which vaporize at even lower temperatures than volatile organics, were also found by the recent CPSC study to be present in the Tennessee homes from the same sources as the VOCs, although they could not be adequately measured using the available equipment. These compounds, the researchers said, may actually be of greater toxicological concern than those easily identified and measured.[11] (A broader discussion of organic chemicals appears in chapter 2, in the sections on Cleaning Products, Hobbies, and Pesticides, along with the accompanying Tables C, G, and I, respectively.)

Concentrations For concentration levels, see Table E.

Adverse health effects Because of the vast number of organic compounds found inside the home, it is virtually impossible to enumerate the potential health effects of the category as a whole. The accompanying table lists some of the chemicals categorized by the CPSC study and notes their health effects (see Table F).

Table F: Adverse Health Effects of Selected Organic Indoor Air Pollutants

Pollutant	Health effects
Benzene	Human carcinogen. Benzene is associated with several forms of leukemia. Chronic and toxic effects on bone marrow result in the reduction of blood clotting abilities and anemia. Acute toxic effects primarily affect the central nervous system. Respiratory irritant.
Toluene	Embryotoxic and teratogenic in mice. Inhalation exposure causes throat and eye irritations and adverse central nervous system effects. May cause anemia.
Ethylbenzene	Causes mucous-membrane irritation in humans. Both inhalation and dermal exposure can cause an irritant response.
Xylene	Embryotoxic. Suspected teratogen in rats and mice. Central nervous system depressant and mucous-membrane irritant. Possibly injurious to heart, liver, kidney, and nervous system.
Cumene	Irritant to mucous membranes. Causes liver, lung, and kidney damage.

Benzaldehyde	Irritant to respiratory system and eyes; chromosomal changes in *in vitro* test systems; sensitizer and central nervous system depressant in humans and animals. Causes kidney and brain damage in animals.
Mesitylene	Mucous membrane irritant; central nervous system depressant.
Limonene	Skin irritant; sensitizer; possible tumor promoter.
Naphthalene	Skin or mucous membrane irritant; vapors may cause headache, loss of appetite, nausea, neuritis (a painful inflammation or lesion of a nerve), kidney damage. May cause excessive destruction of red blood cells.
2-Methylnaphthalene	Adverse effects on liver function in rats. High concentrations can damage the upper respiratory tract.
Diethylphthalate	Causes chromosomal aberration in *in vitro* mammalian tests; fetotoxic and teratogenic in rodents.
Dodecane	Co-mutagen; co-carcinogen (tumor promoter); irritant; narcotic at high concentrations.
Tridecane	Irritant; narcotic at high concentrations; causes asphyxia.
Tetradecane	Co-mutagen; co-carcinogen (tumor promoter); irritant; narcotic at high concentrations.
Pentadecane	Irritant.
Hexadecane	Irritant; causes asphyxia.
Methylene chloride	Mutagenic to bacteria and yeast; induced cell transformations; animal carcinogen; causes central nervous system depression and decreased myocardial contractility. Causes increased carboxyhemoglobin levels in the blood (reducing blood's ability to carry oxygen to the cells).
Perchloroethylene	Animal carcinogen; causes central nervous system depression and liver and kidney damage.
1,2-dichloroethane	Animal carcinogen; possible mutagen; causes central nervous system depression and liver and kidney damage.
Trichloroethane	Causes irregular respiration, semiconciousness (10,000 ppm); reduced activity (5,000 ppm); no other adverse effects.
Trichloroethylene	Animal carcinogen; subject of OSHA carcinogenesis study.
Dichloropropane	Increased ratio of liver weight to body weight and kidney weight to body weight ratios. No other adverse effects.
Trichloropropane	Mutagen.
Carbon tetrachloride	Suspected carcinogen.

Vinyl chloride	Highly toxic; suspected carcinogen.

From U.S. Consumer Products Safety Commission: "Status Report on Indoor Pollution in 40 Tennessee Homes"

Ozone

Ozone (O_3) is virtually unknown in pure, ground-level air. Ambient ozone is a product of the reaction of industrial pollutants, such as nitrogen oxides and hydrocarbons from oil products and solvents, with sunlight, and is one component of smog. Indoor ozone concentrations are largely influenced by outdoor concentrations, with indoor levels typically 40 to 60 percent of the outdoor values. Additional ozone may be produced indoors by photocopiers, laser printers, mercury enhanced lightbulbs, some electrostatic air cleaners and any equipment that uses high voltage or ultraviolet light. Because of this, poorly ventilated offices and especially rooms housing photocopying machines or laser printers may accumulate significant levels of ozone. It is easily recognizable by its strong, pungent odor.

Even if indoor values are only 20 to 80 percent of outdoor levels, it must be realized that indoor ozone exposure (which is concentration multiplied by time) actually makes indoor ozone exposure greater than outdoor exposure for many people.

Concentrations For concentration levels, see Table G.

Adverse health effects Ozone is a pulmonary irritant that affects the mucous membranes, other lung tissue, and respiratory functioning. At 1 mg/m^3 (0.51 ppm), ozone can cause increased susceptibility to respiratory infections. At 2–4 mg/m^3 (1.02–2.04 ppm), it can cause headache, chest pains, and dryness

Table G: Recorded Ozone Levels

In 1972 the Food and Drug Administration established a limit of 0.05 ppm for ozone in houses, apartments, hospitals, and offices; work place limits are 0.1 ppm or 0.2 mg/m^3 averaged over 8 hours; ASHRAE guidelines for indoor air suggest 0.1 mg/m^3 as a continuous limit. In 1979 the EPA adopted a standard of 0.12 ppm in ambient air, not to be exceeded more than once each year. The deadline for reaching that standard is Dec. 31, 1987, although total compliance is not expected to be reached by that date, especially in major metropolitan areas. Despite this, there is now pressure on the Federal government to lower this standard to as little as 0.08 ppm based on the most recent scientific studies of ozone and its effect on human health.

Location	Level	Source
Photocopying machine room	<0.002-0.068 ppm (after 40 mins.-2 hrs of operation)	(1)
Photocopying room at operator level	4-300 micrograms/m^3 (depending on servicing of machine)	(2)
Iowa State University energy research house	0-20 micrograms/m^3 (0-8ppb) (range most frequently observed)	(2)
17 residences and buildings	0.002-0.012 mg/m^3 (typical 1-hour averages)	(3)
12 residences	0.28 mg/m^3 (maximum 1-hour average)	(3)
Jumbo jets and supersonic aircraft traveling near earth's ozone layer	<0.02 ppm	(2)

Sources: (1) Wadden, R.A., and Scheff, P.A.: *Indoor Air Pollution*, p. 71.
(2) Meyer, Beat: *Indoor Air Quality*, p. 176.
(3) Sandia National Laboratories: *Indoor Air Quality Handbook for Designers, Builders, and Users of Energy-Efficient Residences*, p.67.

in the upper respiratory tract.[12] Because the effects of ozone are increased synergistically by low humidity and the presence of such other pollutants as cigarette smoke, a combination of conditions commonly encountered in commercial aircraft, the Federal Aviation Administration has done several studies on the health effects of ozone. According to these, short term exposure to levels up to about 0.3 ppm (experienced in commercial air flights) generally results in noticeable respiratory discomfort but no long term changes in respiratory function.[13] More recent studies, however, indicate that adverse health effects on humans may begin at concentrations of 0.12 ppm and possibly even below, and that lasting damage can result from repeated exposure to elevated ozone concentrations.

A study by New York University showed decreased lung function in children exposed to as little as 0.113 ppm and suggested that the loss of lung function may be due to cumulative exposure to ozone rather than short-term exposures during peak periods. A Harvard University study indicated an association with decreased lung function in children exposed to as little as 0.078 ppm ozone. Despite National Ambient Air Quality Standards for ozone of 0.12 ppm, more than half of the U.S. population resides in areas unable to attain compliance.

Particulates

Particulates are not a single type of pollutant, but a description of the physical state of many

Eyes, skin, and respiratory tract are targets

pollutants—that is, all suspended solid or liquid particles (such as dust or sneeze droplets) less than a few hundred micrometers in size. (Paper, for a reference, is about 100 micrometers thick.) This definition applies to all particulates—referred to as the total suspended particulates (TSP)—in the environment. The effect that particles have on health depends on their chemical makeup and where in the body they come to lodge. The eyes and the skin are easy targets for all particulates, but the respiratory tract, which is the area of primary concern, is exposed to only a small part of the TSP, known as respirable suspended particles (RSP).

Particles larger than about 50 micrometers settle out of the air quickly (although they can easily be resuspended). Most of those smaller than 0.05 micrometers are too fine to be retained and are exhaled, unless they dissolve or react with tissue they may meet. Of the remaining particles, those larger than about 15 micrometers are filtered out by the nose; those smaller than 15 micrometers move into the nasopharyngeal tract (nose, sinuses, and throat), where a large fraction of the larger ones lodge and are either transferred to the gastrointestinal tract or eliminated with sputum; those between 0.1 and 10 micrometers can be absorbed into the tracheobronchial tract or move into the pulmonary cavity where they can be retained for months, or even years, as in the case of asbestos fibers. This last size category is considered the most serious in terms of long-range or chronic health effects. Particles smaller than five micrometers have a high likelihood of being deposited in the sensitive bronchioles and alveolar spaces in the lungs, and of being transferred to the blood or lymph, which can transport them anywhere in the body.[14]

Among the pollutants that appear in the form of particulates are asbestos and other fibrous building materials, radon progeny, smoke, organic compounds, infectious agents (bacteria, viruses, and fungi), allergens, and heavy metals, such as cadmium (found in cigarette smoke).

Concentrations

Clean ambient air contains levels of particulates in the range of 20–40 micrograms per cubic meter, in the form of sea spray, soluble salts, organic material, and microbes. The additional heavy load of particulates we often encounter outdoors is attributable to such sources as industrial smoke, vehicular emissions, road dust, pollen, and pesticides. Because most of these are generated outdoors, indoor air usually has a lower particulate concentration.

The most important sources of indoor particulates are windblown dust, house dust, and tobacco smoke. This last source is so powerful a generator of particles that the particulate content of indoor air may become higher than ambient levels in a poorly ventilated space with several heavy smokers (see Tables O, P, and Q). Wood combustion can also raise indoor particulates dramatically, both from backdrafts directly into the house and from windblown smoke from outside. People themselves produce half a million particles per minute in sizes larger than 0.3 micrometers, making people a serious source of biological pollutants in crowded environments such as office buildings or indoor sports centers.

Adverse health effects

Because of the diversity of particulates and of their chemical nature, it is not practical to consider adverse health effects for this category as a whole. Individual adverse health effects of many of the pollutants encountered as particulates are described separately in this compendium. It should be noted, however, that the lungs have very efficient mechanisms for cleansing themselves of much of the particulate matter they absorb, usually within a day after deposition.

Pesticides

Several hundred pesticides, meant to kill insects, rodents, and other pests, are currently available to homeowners and exterminators. Because of the scope of the subject, only a few of the more common ones can be discussed here.

Concentrations

For concentration levels, see Table H.

Adverse health effects

Many of the pesticide formulas contain chemicals that are fat soluble and can build up in human adipose (fatty) tissue. Some have a lung tissue half-life that could represent a long term hazard. In some cases, a safe threshold limit for concentration in human beings has not decisively been established. Although additional research is needed, it is believed that the polyhalogenated hydrocarbons used in many insecticides may increase

Table H: Airborne Concentration of Seven Pesticides During Day of Application and Three Days Following

	Concentration (micrograms/m³)			
Insecticide	**Day of Application**	**Day 1**	**Day 2**	**Day 3**
Acephate	1.3	2.9	0.5	0.3
Bendiocarb	7.7	1.3	—	—
Carbaryl	1.3	0.2	0.1	0.01
Chlorpyrifos	1.1	1.1	0.8	0.3
Diazinon	1.6	0.6	0.5	0.4
Fenitrothion	3.3	1.1	0.8	0.5
Propoxur	15.4	2.7	1.8	0.7

Source: Meyer, Beat: *Indoor Air Quality* p.210.

Note: Even after airborne concentrations have declined, residue of the pesticides in the form of dust may be resuspended at any time.

the rise of coronary heart disease by binding to estrogen receptors and increasing cholesterol and triglyceride levels in the body (see Table I).

Polychlorinated Biphenyls (PCBs)

Used for almost 50 years

Polychlorinated biphenyls (PCBs), organic compounds that are nonflammable, chemically and thermally stable, and make excellent electrical insulators, were in great use in the United States for almost 50 years. They were used in such diverse products as fluorescent light ballasts (starters), small electrical capacitors, textile dyes, printing inks, paints, carbonless copy paper, and fireproofing agents. Beginning in the mid-1960s, PCBs were discovered to be environmental contaminants residing everywhere from human

Table I: Pesticides and Their Potential Health Effects

Pesticide	Target Use	Potential Adverse Health Effects
Aldrin	Ants, termites	Suspected carcinogen
Dieldrin	Ants, termites	Suspected carcinogen
Heptachlor	Ants, termites	Suspected carcinogen
Lindane	Ants, termites	Suspected carcinogen
Chlordane	Ants, termites	Highly toxic; persistent in soil, timber, and human tissues; animal carcinogen
Mirex	Ants, cockroaches	Oxidizes to form Kepone
Kepone	Ants, cockroaches	Causes cancer in mice; adversely affects reproduction in lab animals; highly persistent; accumulates in human tissue; several workers in a kepone plant suffered severe illness
Diazinon	Ants, cockroaches	Highly toxic to rats
Malathion	General (insects)	Low mammal toxicity; readily degraded
Dichlorvos	Flea collars; fly strips (Vapona)	Can cause hypersensitivity; under investigation for oncogenic, mutagenic, fetotoxic, and neurotoxic effects
Propoxur	Flies, mosquitos, ants, cockroaches	Unknown
Pentachlorophenol (PCP)	Fungicide; wood preservative	Highly toxic to humans; contains traces of dioxin, which is also highly toxic; persists in air of log cabins. In 1980 the blood serum of 30 log cabin dwellers had a level of PCP seven times higher than a control group.

Source: Meyer, Beat: *Indoor Air Quality*, pp. 119-126.

mother's milk to Antarctic snow to Icelandic plants. Although they ceased to be produced in 1977 and have been phased out of use in virtually all applications, their extreme stability causes them to remain a polluting presence around the globe.[15]

The ballasts of fluorescent light fixtures manufactured before 1978 contain PCBs that may be released as a vapor when these lamps catch fire or explode. Defective light ballasts may emit PCBs constantly.

Concentrations

In 1980, Kathryn MacLeod of the U.S. Environmental Protection Agency conducted research to determine the levels of PCBs encountered in residences and workplaces (see Tables K and J). Of the nine houses, four had pre-1972 fluorescent lighting fixtures in their kitchens.

Her results indicated that the average level of PCBs in the offices of the industrial research building was five times higher than the level outdoors, and the level in the laboratories was 10 times higher than ambient concentrations. The air outside one home showed a PCB level of 4 ng/m^3, while indoor concentrations were as high as 500 ng/m^3 in the fluorescent-lit kitchen. In all cases, the highest PCB levels were found in the kitchen. Although MacLeod found no clear correlation between PCB concentration and the type of lighting used, she believed that PCB emissions from other electrical appliances in the kitchen were to blame for the consistently high levels.

Fluorescent light ballasts

While testing for PCBs in the work place, a fluorescent light ballast burned out, allowing her to examine the persistence of PCBs in the air. In the room containing the burned-out light ballast, air levels of PCBs on the day of burnout were found to be over 50 times higher than normal for that room, and levels remained elevated for three to four months afterward.[16]

Table J: Airborne PCB in Homes

Site	Month sampled	Average concentration (ng/m^3)	Fluorescent lighting
Kitchen A	June	480	
Kitchen B	—	180	
Kitchen C	June, Nov.	250	yes
Kitchen D	April	210	yes
Kitchen E	July, Dec.	240	
Kitchen F	June, July, Oct., Nov.	150	
Kitchen G	July, Sept.	580	yes
Kitchen H	March, April, Nov.	260	
Kitchen I	March, April	500	yes
Living room D	April	39	
Bedroom E	July, Dec.	170	
Basement E	July, Dec.	120	
Garage F	June, July, Oct.	64	
Library I	March, April	400	
Outdoors	March	4	

One nanogram (ng) equals one billionth of a gram

Source: MacLeod, Kathryn E.: "Polychlorinated Biphenyls in Indoor Air," *Environmental Science & Technology* (August, 1981) p. 927.

Adverse health effects

PCBs are highly toxic substances, especially when ingested. This was first discovered in 1964 when mass poisoning in Japan was traced to rice oil contaminated by PCBs emanating from a heat exchanger. In addition, PCBs are highly suspected carcinogens. Although additional study is needed to link PCBs with cancer in humans, animal studies have found the compound to be carcinogenic, teratogenic, mutagenic (causing cancer, fetal death, or mutation), and to have

Table K: Airborne PCB Levels in Work Places

Site	Period of sampling	Average concentrations (ng/m^3)
(Industrial Research Facility)		
Laboratory A	6 months	200
Laboratory B	2.5 months	240
Office A	1 week	110
Office B	2 days	80
Outdoors	1 week	18
(Academic Facility)		
Laboratory	2 days	210
Outdoors	2 days	4
(Shopping Complex)		
Office	2 days	44

One nanogram equals one billionth of a gram

Source: Macleod, Kathryn E.: "Polychlorinated Biphenyls in Indoor Air," *Environmental Science & Technology*, August, 1981, p. 927.

developmental effects on laboratory animals. PCBs can build up in the human body, prolonging the dose beyond the actual time of exposure and crossing the placenta to the fetus in pregnant women.

Radon

In 1977, the United Nations Committee on the Effects of Atomic Radiation (UNSCEAR) estimated that radon gas is responsible for almost one third of the radiation exposure that the average person receives each year. Radon is a naturally occurring radioactive gas found in low concentrations almost everywhere on earth. It is

Radon chain

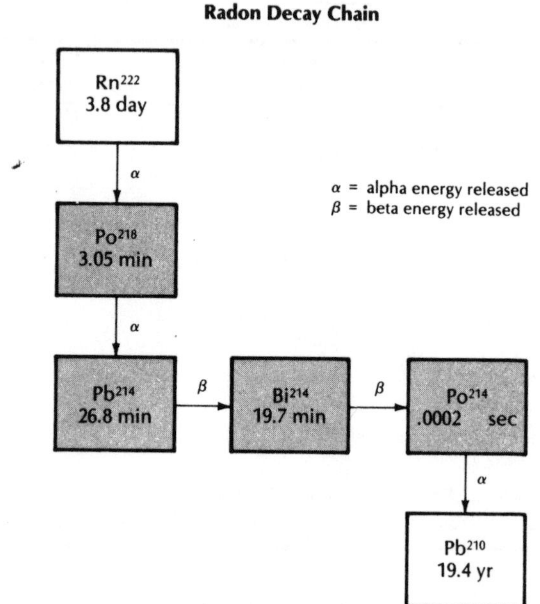

Note: Shaded isotopes are of primary radiological concern because of possible inhalation and subsequent alpha decay in the lungs.

Figure I.

colorless, odorless, water soluble, and chemically inert. It is produced as part of the radioactive decay chain of the element uranium-238 (see accompanying diagram) and further decays into radioactive progeny (sometimes referred to as "daughters"), releasing alpha, beta, and gamma radiation along with each change. The amount of time it takes for half the atoms of each element to decay is called the half-life, which is noted on the diagram between each element. For our purposes, the chain ends with lead 210, which has an extremely long half-life—that is, it does not decay further for a period of time long enough to render it fairly harmless in the home. It is radon's decay and that of its progeny that pose the threat, since they rapidly release doses of radioactivity in the presence of human cells.

In well water and soil

A significant amount of radon is present in well water and soil in many areas of this country, including much of Maine, Florida, Colorado, Pennsylvania, New Jersey, New York, Illinois, Montana, and Maryland. It is commonly associated with granitic bedrock and can also be present in natural gas and coal deposits found in this rock. In its naturally gaseous state, radon can rise easily through airspaces in the soil to enter the house through its basement, be released from agitated or boiled water, or escape during use of natural gas. Once indoors, the radon progeny show a magnetic affinity for airborne particulates such as dust and tobacco smoke, hitchhiking on these into the lungs of residents. Many of the progeny end up attached to solid surfaces, however, which renders them relatively harmless unless they are reintroduced into the air by means such as dusting or sweeping.

Radon also enters our homes by its presence in many earth-based building materials, such as stone, brick, concrete, and cement. In some cases these materials were made from the waste products of phosphate or other mining operations, or simply from rock high in radon content (see Table L).

Table L: Approximate Contributions of Various Sources to Observed Average Indoor Radon Concentrations

Source	Single-family houses (pCi/L)	Apartments (pCi/L)
Soil	1.5	<1
Water (public supplies[a])	0.01	0.01
Building materials	0.05	0.1[b]
Outdoor air	0.2	0.2
Observed indoor concentration	1.5	0.3

Source: Nero, Anthony V.,Jr.: "Indoor Concentrations of Radon-222 and its Daughters: Sources, Range, and Environmental Influences," *Indoor Air and Human Health*, Gammage, R.B., and Kaye, S.V. eds., p. 54.

[a] Applies to 80% of population served by such supplies; contribution from water may average about 0.4 pCi/l in homes using private wells, with even higher contributions in high-activity areas.

[b] A higher concentration in apartment air is suggested on the presumption that on the average, apartments have a higher amount of radon-bearing building materials per unit volume than do single-family houses.

Concentrations

Radioactivity has a complex measurement system all its own. The Curie (Ci) is a commonly used measurement of radioactivity, but it is too large to use in relation to the amounts discussed in the case of radon. (In scientific terms, it equals the disintegration of 37 billion nuclei per second.) Therefore, the usual unit used with respect to radon is the picoCurie (pCi), equal to one-millionth (.000001) Curie. In measuring radon concentration, the unit used is picoCuries per liter (pCi/l) of air or water.

Table M: Representative Exposures to Radon Progeny

Subjects or Location	WL	WLM (Lifetime)
Uranium miners*	1-20	100-10,000
Outdoors	<0.001	—
Indoors	<0.01	10

*Includes exposure before mid-1960s.

WL = Working level of potential alpha energy
WLM = Working level months

Source: National Research Council: *Indoor Pollutants,* p. VII 9

Table N: Selected Radon and Radon Progeny Levels in U.S. Residences (Annual average indoor air quality guideline is 0.01 WL)

Location[a]	Radon concentration[b] pCi/m³	Progeny concentration[b] WL	Comments
Ordinary areas:			
Tennessee	—	0.008 (0.0008-0.03)	Shale area; mostly concrete construction
Boston	0.07 (0.005-0.2)	(up to 0.002)	Single family; 1-6 air changes/hour
New York/New Jersey	0.8[c] (0.3-3.1)	0.004[c] (0.002-0.013)	17 single family; 3 multifamily; 1 apartment building
Illinois	(0.3-33)	—	Wood frame construction; unpaved crawl spaces (windows closed)
San Francisco area	(0.4-0.8)	—	0.02-1.0 air changes/hour (windows closed)
U.S./Canada	(0.6-22)	—	Energy-efficient houses; 0.04-1.0 air changes/hour (windows closed)

Special areas:			
Grand Junction, Colorado	—	0.0006[c]	Control for remedial action program (which has included houses in 0.02-1 WL range)
Florida	—	0.004 (0.0007-0.014)	Controls on unmineralized soils
	—	0.014	Houses on reclaimed phosphate lands
Butte, Montana	—	0.02	Intensive mining area
Anaconda, Montana	—	0.013	Intensive mining area

WL = Working level

Source: National Research Council: *Indoor Pollutants*, pp. IV 15, IV 16

[a]Single-family residences except where noted.

[b]Averages; values in parentheses are ranges. All measurements are in living space; values in basements are typically higher.

[c]Geometric mean.

Alpha particles Radon progeny are measured in "working levels" (WL) of potential alpha energy concentrations. Technically, these are defined as any combination of radon progeny in one liter of air whose radioactive decay will result in the release of 1.3×10^5 megavolts (MeV) of alpha-particle energy (130,000 million electron volts). The alpha-particle itself is a positively charged particle (a helium nucleus) emitted in the decay process. This measurement is clearly beyond the scope of most of the nonscientific community but becomes more easily comprehended when looked at in a comparative manner in the accompanying tables (see Tables M and N).

Adverse health effects

An understanding of just what happens when radon and its progeny enter the body through inhalation or ingestion is essential to a clear picture of the element's adverse health effects. Radon progeny ingested with water cannot pass through the intestinal walls into the blood, although radon itself (a gas) can diffuse through these walls. The progeny of this radon can affect the stomach and other organs, although the amount of these progeny is generally considered to be too insignificant to have an effect.

Because radon is chemically inert (it does not readily react with other elements), radon that is inhaled is also exhaled with few or no consequences. The radon progeny, however, are chemically active metals (lead, bismuth, polonium), and their charges make them adhere to any particles or surfaces with which they come into contact. When inhaled, almost one third of the progeny come into contact with tissue in the lungs and stick there. Radiation from these progeny can affect the molecules in the surrounding cells.

The emission of alpha particles by the radon progeny is the major source of health concern. An alpha particle (really a helium atom moving at high speed) collides with a great many molecules, knocking electrons off of these molecules as it goes, in a process known as "ionization." In air, the initial energy of the alpha particle carries it about two to three inches. In solid matter, such as body tissue, the range is much shorter—about one-thousandth of an inch—because of the greater density of the molecules through which it is passing.

Cancer and radon

The ionization of the molecules can cause them to be broken and changed, often in ways harmful to living tissue. For example, if the information in a genetic molecule such as DNA is changed, the functioning of a whole cell can be altered, particularly when it reproduces itself. This, for example, could cause a normal healthy cell to become cancerous. If enough cells become damaged, the tissue, and finally the organ itself, becomes disordered—that is, cancerous. There are no early warning signs that cancer may develop a decade or two after exposure. Radiation can also disrupt enough chemical bonds in a chromosome to cause mutation.

The other forms of radiation emitted by radon progeny—beta and gamma radiation—can penetrate deeper into tissue than can alpha radiation, but the density of ionization from beta and gamma radiation is much less than from alpha. In other words, an alpha particle passing through a large DNA molecule in a cell nucleus may break that molecule in three or four places, whereas beta or gamma radiation might break it in only one place or even in none at all. Cells are able to repair themselves more readily from the single break of beta or gamma radiation.[17]

Because the adverse health effects of low-level radiation may take years or even decades to develop, much of the body of knowledge about the potential adverse health effects of radon and radon progeny is extrapolated from and modeled on the effects of similar radiation on uranium miners. True, these miners were exposed to much higher doses of radiation, but the time periods for which they were exposed were much shorter, and the exposed persons were strong, generally healthy men rather than the entire population, which contains a significant portion of more susceptible people, such as infants, children, the ill, and the elderly (see Table M).

In assessing the effects of low-level radiation, the commonly held assumption is that the incidence of lung cancer is always proportional to the cumulative dose of radiation. That is, there is no smallest or threshold dose, and doubling the lifelong exposure doubles the incidence of disease.

Latent development

A high percentage of miners exposed to radon gas died of lung and bronchial cancers, and many of those exposed to sublethal doses contracted kidney or spleen lesions. In most cases there was a delay of from 10 to 25 years between the time of exposure and the appearance of lung cancer. Miners today are still dying from lung cancer caused by radiation exposure in the 1950s.

Almost all the affected miners were exposed to an average 500 rems of radiation. This would be equivalent to living in a house with airborne radon levels of 10 pCi/l for 60 years. The lowest radiation dosage for which there is conclusive evidence of higher than normal incidence of lung cancer among miners is around 40 to 50 rems, equal to living in a house with a radon level of 2 pCi/l for 20 years.

Based on the data, there is no doubt that radon

and its progeny in sufficient doses can produce lung cancer in people. There is also evidence that smoking increases the risk of radon-induced cancer, although additional research is needed to quantify the increased risk. As for the risk assessment of increased lung cancer from indoor radon, the EPA has claimed that as many as 20,000 lung cancer deaths each year may be the result of exposure to radon and its progeny. The National Research Council has noted that data indicate that we can expect 2.2 to 10 cases of lung cancer from radon exposure per million persons per year per working-level month (WLM). Using the exposures given in Table N, the entire U.S. population of over 200 million people can expect about 10,000 cases of lung cancer per year from radon exposure. If this is so, of the approximately 100,000 new cases of lung cancer diagnosed annually, some 10 percent can be attributed to radon. Another 80 percent are thought to be caused by cigarette smoking. Although additional study is necessary to give precise figures in this area, there is no doubt that the threat of serious, harmful effects from radon and radon progeny must be reckoned with.

Sulfur Dioxide

Killer smogs

During the heyday of coal combustion, sulfur dioxide (SO_2) was the most prevalent and noxious air pollutant in industrial and urban centers. It was responsible for the killer smogs that plagued London, Belgium, and Appalachia, among other areas. Since the burning of coal has dropped off dramatically, the main source of SO_2 is coal combustion in electric power plants, oil refineries, and some industrial heating systems.

After the sulfur dioxide leaves the stacks, it can remain windborne for many hundreds of miles, reacting along the way to produce sulfate (SO_4) particles. These further react with water droplets and water vapor to produce sulfuric acid (H_2SO_4), which is diluted in rain to create "acid rain." This acid rain raises the acidity of the soil and water into which it falls to the point where vegetation and marine wildlife can no longer survive. Millions of acres of forestland and many rivers and streams are damaged or even destroyed annually by acid rain.

Concentration

Indoor levels of SO_2 are generally lower than outdoors because SO_2 is readily absorbed by most building materials. There are virtually no major sources of SO_2 indoors unless coal is used as a heating material in a fireplace or stove. In that case, SO_2 levels have been measured at 0.4 ppm at night and 0.8 ppm during the day. In noncoal-burning residences, levels of 3 to 29 parts per billion were recorded in a mobile home in Baltimore; 0 to 110 ppb in an apartment house in that city; and 1.4 to 2.5 ppb in a residence in Washington.

Adverse health effects

Although soluble sulfate is metabolized by the body, apparently without ill effect, respirable sulfate particles can cause asthma and other respiratory problems. Sulfur dioxide is highly irritating to eyes, skin, and mucous membranes.

Tobacco Smoke

Mixture of over 3000 chemical compounds

Tobacco smoke is a complex mixture of more than 3,000 chemical compounds—including gases, aerosols, respirable particles, dust, and smoke—many of which are toxic, carcinogenic, or even radioactive. (See chapter 2 for a list of these components.) Continuing research has given rise to the division of cigarette smoke into two categories: mainstream smoke, which is inhaled by the smoker; and sidestream smoke, which rises directly from the burning tobacco. Sidestream smoke often contains concentrations of pollutants higher than those in mainstream smoke. This is true for two reasons: First, because the cigarette burns hotter when the smoker inhales, thus attaining more complete combustion and yielding fewer pollutants, and second, because the smoke is filtered somewhat by the tobacco in the cigarette itself and any filter the cigarette may have, before being inhaled by the smoker. The smoker's lungs then act as a very efficient filter, trapping up to 90 percent of the rest of the pollutants before the smoke is exhaled. In addition, sidestream smoke is produced during about 96 percent of the total burn time, whereas the smoker only inhales about 8 to 10 times. In enclosed areas, such as homes, vehicles, sports arenas, public buildings, and stores, involuntary (or passive) smokers may become victims of the same significant health threats as those who choose to smoke.

The actual dose of contaminants that a nonsmoker receives, however, is dependent on proximity to the burning cigarette and the quality of ventilation in the room. In no way does the passive smoker receive a lung dose equal to what the smoker receives for each cigarette consumed. Still, in buildings with extensive air-recycling equipment (heaters, air conditioners, humidifiers), one component of cigarette smoke—carbon monoxide—presents a special problem. Because it is a gas, it remains suspended in the air indefinitely and is not removed by most standard filtration systems. The only way of reducing CO concentration is to replace polluted air with fresh air. In energy-efficient buildings the air tends to be self-contained and the CO levels in some buildings have been measured in excess of the National Ambient Air Quality Standard of 9 ppm. In the presence of heavy smoking and in poorly ventilated rooms, concentrations have even exceeded the EPA ambient air standard of 35 ppm averaged over an eight-hour period.

Concentrations For levels of concentration, see Tables O, P, and Q.

Table O: Measurement of Selected Components of Tobacco Smoke Under Natural Conditions

Location	Component	Smoking section	Nonsmoking control section (where given)
	Nicotine		
Commuter train		0.0049 mg/m^3	
Commuter bus		0.0063 mg/m^3	
Bus waiting room		0.001 mg/m^3	
Airline waiting room		0.0031 mg/m^3	

Restaurant		0.0052 mg/m^3	
Cocktail lounge		0.0103 mg/m^3	
Student lounge		0.0028 mg/m^3	
Submarine (157 cigarettes/day)		0.032 mg/m^3	
Submarine (94-103 cigarettes/day)		0.015-0.035 mg/m^3	
	Particles		
House (1 cigarette)		$48 \times 10^6/ft.^3$	$.9 \times 10^6/ft.^3$
Airplane flights		$<.120$ mg/m^3	
Arenas		.367 mg/m^3	.068 mg/m^3
Tavern 1 (6 air changes/hour)		.33 mg/m^3	
Tavern 2 (1-2 air changes/hour)		.98 mg/m^3	
	Carbon monoxide		
Tavern l (6 ach)		12.5 ppm	
Tavern 2 (1-2 ach)		17 ppm	
Restaurants		8-28 ppm	
Arenas		14.3 ppm	3 ppm
Offices		2.7 ppm	
Conference room (8 ach)		8 ppm	1-2 ppm
Conference room (6 ach)		10 ppm	1-2 ppm
Submarines		<40 ppm	
Intercity bus (15 ach):			
23 cigarettes		33 ppm	
3 cigarettes		18 ppm	
Automobile four cigarettes:			
(35 km/hr, no ventilation)		24.3 ppm	
(80 km/hr, no ventilation)		12.1 ppm	
(30 km/hr, air vents open)		15.7 ppm	
(3 km/hr, vents open, fan on)		12.0 ppm	
Theater		3.4 +/− 0.8 ppm	1.4 +/− 0.8 ppm
Room (78.3 m^3): 3 smokers		15.6 ppm	

	Dimethylnitrosamine	
Train (bar car)		.13 ng/l
Bar		.24 ng/l
	Benzo(a)pyrene	
Restaurant		.0002-.0046 mg/m^3
Room*		1-5 ng/m^3

* From Meyer, Beat: *Indoor Air Quality*.

Source: National Research Council: *Indoor Pollutants*, pp. VI 103-106

ach = air changes per hour

Adverse health effects

Many of the individual components of tobacco smoke are given separate listings within this compendium of pollutants and their health effects. The subject of tobacco smoke as a whole is worthy of its own listing, however, because as in so many cases, "the whole is greater than the sum of its parts"—that is, the synergy among the components tends to exacerbate the effects of any of them alone.

Responsible for many cardio-pulmonary ailments

Extensive research over the past 20 years has resulted in the definitive warning that voluntary smoking is hazardous to human health, often resulting in heart disease, lung cancer, emphysema, and other cardiopulmonary ailments. Tobacco smoking is believed to cause the death of more than 50,000 Americans each year—due mainly to lung cancer, oral cancer, and cancer of the larynx and esophagus. However, the health effects of passive smoking, or environmental tobacco smoking (ETS), are only now coming to light. Draft findings of a 1990 EPA report reviewing more than 20 studies published since the 1986

Surgeon General's Report that first stated the harms of ETS, found that passive smoking causes 3,800 lung cancer-related deaths each year. Other studies note that passive smoking slows down the heart's ability to get oxygen and contributes to blood clotting, causing a 30 percent increase in risk of death from heart attacks by non-smokers who live with smokers.

At the simplest level, tobacco smoke contains many substances that irritate the eyes, nose, and throat, and result in eye irritations, headache, cough, sneezing, sore throat, and nasal discomfort. In addition, the odor is considered significantly unpleasant by a large segment of the population.

Beyond these temporary conditions, however, there is strong evidence that involuntary respiration of tobacco smoke can produce chronic adverse health effects in children, the elderly, people with existing ill health, and even otherwise healthy adults. In one Japanese study by Hirayama, more than 91,000 nonsmoking Japanese housewives were studied from 1966 to 1979. During this time, 174 of the women died of lung cancer. Hirayama found that in urban areas, the nonsmoking wives of men who smoked more than 20 cigarettes per day had 2.4 times the risk of incurring lung cancer than did the wives of non-smokers. In rural areas, the risk increased to 4.6 times. He attributed the lower perceived risk in urban centers to the greater presence of other environmental carcinogens, which made it difficult to detect the effect of cigarette smoke alone.

Table P: Indoor Levels of Respirable Suspended Particles (RSP)

Location	Number of occupants	Number of smokers	Indoor level*	Outdoor level*
Cocktail party	14	2	350	—
Lodge hall	350	40	700	60
Bar and grill	75	9	590	63
Pizzeria	50	5	415	40
Church: Bingo game	150	20	280	—
Services	300	0	30	—
Bowling alley	128	14	200	50
Hospital waiting room	19	2	190	28

* All levels are given in micrograms per cubic meter.

Source: Lowery, A.H. and J.L. Repace, "Indoor Pollution, Tobacco Smoke, and Public Health," *Science*, vol. 208, (May, 1980) in Turiel, Isaac: *Indoor Air Quality and Human Health*, p. 73.

Table Q: Concentration of Respirable Suspended Particles in the Home

Site description	Level (micrograms/cubic meter)
Home without smokers	15 +/− 2
Home with one smoker	35+/− 8
Home with more than one smoker	50 +/− 15

Source: Meyer, Beat: *Indoor Air Quality*, p. 181.

Similar results were found in a Greek study.[18]

Upper respiratory tract infections in children

In the many worldwide studies of children and smoking, almost all the evidence points to increased incidence of upper respiratory tract infections, bronchitis, asthma, and pneumonia in children of parents who smoke. The relationship is believed to have two causes: First, the direct effect of the tobacco smoke on the children, who are particularly sensitive to the effects of pollution because of their greater ratio of breathing to body weight; and second because the smoking parents are sick more often themselves and are subsequently infecting their children.

A Finnish study of 12,000 children under the age of five found that those with mothers who smoked had significantly higher incidence of illness (due mainly to respiratory disease), were more likely to be hospitalized, and that their hospitalizations were longer than children of nonsmoking mothers. The most pronounced effect was in the first year of life.[19]

A British study of 2,205 children over the first five years of their lives also found a relationship between the smoking habits of both parents and the incidence of pneumonia and bronchitis. The relationship was greatest when both parents smoked, and it increased with the number of cigarettes smoked each day. This study also found a strong relationship between cough and phlegm production in parents and respiratory infections in children.[20] In a study of children aged five to nine years old, parental cigarette smoking was related to the occurrence of a persistent wheeze and decreased respiratory functioning in children.[21]

Speaking on behalf of the American Academy of Pediatrics, Dr. Jonathan D. Klein told the Senate Subcommittee on Finance in May 1990 that children raised by smokers face an increased risk of developing a range of health problems, including more frequent lower respiratory tract infections, otits media (ear infections), pneumonia and potentially fatal syncytial virus. He also cited long term problems including decreased lung function, less lung growth and higher rates of asthma than children of non-smokers. The early exposure to tobacco smoke is also believed to increase the risk of lung cancer and heart disease in children, since children exposed to passive smoking from birth had higher cholesterol levels and lower levels of HDL, a blood protein believed to help prevent heart attacks.

A clear danger

Among patients with coronary heart disease, angina pectoris, and lung diseases, the effects of sidestream cigarette smoke are also clear. Along with elevated resting heart rate, blood pressure, and carboxyhemoglobin levels, there is a significant reduction in ability to perform physical exercise before chest pain develops. The body of literature on involuntary smoking has persuaded the National Academy of Sciences to conclude that "Public policy should clearly articulate that involuntary exposure to tobacco smoke has adverse health effects and ought to be minimized or avoided where possible."

Table R: Selected Work Place Safety and Health Standards

Pollutant	Concentration[a] (ppm)	mg/m^3
Ammonia	50	35
Carbon dioxide (CO_2)	5,000	9,000
Cresol	50	55
Formaldehyde	2	3
Nitric oxide (NO)	25	30
Nitrogen dioxide (NO_2)	5	9
Ozone (O_3)	0.1	0.2
Propane	1,000	1,800
Sulfur dioxide (SO_2)	5	13

Trichloromethane	50	240
Respirable dust	--	5
Asbestos	Fewer than two fibers	>5 micrograms/cc

[a]Applicable to an 8-hour time-weighted average, except for SO_2 which is a ceiling value.

Source: National Institute of Occupational Safety and Health, as cited in Meyer, Beat: *Indoor Air Quality*

7 Indoor Pollution Redux

Changes in the 90s

In the few years that have elapsed since the first edition of this book appeared in print, massive changes have taken place in the way people look at their indoor environment, the way the government and the courts view the health risks associated with indoor air, and the way the construction and ventilation industries approach their business.

In 1986 the very idea of "indoor pollution" was virtually unheard of. The basic premise of this book was met with stares of incredulity and even derision. Radon, VOCs, combustion byproducts, particulates, all were just so much scientific jargon to most of the public. A small number of people were mildly aware of asbestos, formaldehyde, carbon monoxide, and of course non-smokers were painfully aware of the tobacco issue, but most people were hardly concerned about the effects indoor air quality could be having on their health or that of their families.

Outbreaks of Legionnaires disease did serve to illustrate some immediate manifestations of inadequate or poorly maintained ventilation in large buildings, but most of the public remained sanguine after the news reports faded away.

Since that time, however, much of the information the government and private researchers were unearthing about the indoor environment has been percolating to the surface. Newspapers and magazines began featuring reports and human interest stories. Television and radio jumped aboard the bandwagon, particularly when the Washington offices of the EPA themselves became the site of Building Related Illness. And as the word spread that here was a health risk we had not only to be aware of but act on, people began putting pressure on the government to become more affirmatively involved.

Accordingly, the government, under Title IV of the Superfund Amendments and Reauthorization Act of 1986 (SARA) charged the EPA with establishing a research program regarding indoor air quality, and to disseminate relevant information to the public. EPA also had to submit a report to Congress describing these activities and making recommendations as appropriate.

New EPA findings

Two reports released by EPA in 1987 and 1989 concluded that radon and indoor air pollution pose greater human health risks than virtually any other environmental hazard, and the agency gave them the highest health risk ranking in all three of the regions studied (i.e. New England, the Middle Atlantic and the Pacific Northwest). Following up on their findings that the majority of their funds are being used to address problems that do not pose the greatest risks, these regions have drafted "strategic plans" for fiscal 1992 that will allow them to shift funds from other Superfund areas to indoor air programs.

In order to disseminate information to the public,

EPA fact sheets and booklets

EPA began producing a series of fact sheets called *Indoor Air Facts*, dealing with many aspects of air quality in homes and office buildings. The agency also produced an *Indoor Air Reference Bibliography, Directory of State Indoor Air Contacts, Survey of Indoor Air Quality Mitigation Firms, The Inside Story: A Guide to Indoor Air Quality*, a consumer-oriented booklet, and several publications on radon.

Publications to be produced in 1991 include a physicians' handbook on IAQ and related illness, a homebuilders' guide to IAQ, a manual and brochure on IAQ in school, an IAQ building design text, and even a video on home indoor air quality. These and other booklets are available by contacting the EPA at either its national or regional offices.

Simultaneously, a number of other federal agencies became actively involved in indoor air quality issues, all forming the interagency Committee on Indoor Air Quality (CIAQ).

The Consumer Product Safety Commission (CPSC), under the federal Hazardous Substances Act, has jurisdiction over the products and appliances and many of the building materials contributing to indoor pollution. The CPSC issued regulations or voluntary standards regarding such products as urea-formaldehyde foam insulation, formaldehyde in pressed-wood products, unvented combustion appliances (e.g. kerosene heaters), asbestos-containing patching compounds and other consumer products, humidifiers, vaporizers, air conditioners, coal- and wood-burning stoves and gas stoves.

The Department of Energy (DOE) has long played a major role in indoor air quality research. Its policy goals are: i) elimination of potential hazards to the public and environment from radioactive contamination remaining at facilities and sites used in the nation's atomic energy programs; and ii) development of information to ensure the maintenance of healthful indoor environments with continuing use of energy conservation measures in buildings.

Government agencies increase research

The Department of Health and Human Services (DHHS) and National Institute for Occupational Safety and Health (NIOSH) are the primary agencies conducting building investigations. Other DHHS agencies participating in air quality research or mitigation activities include the Center for Disease Control; National Institute of Health; Health Resources and Services Administration; and the Agency for Toxic Substances and Disease Registries.

A number of other government and quasi-government agencies conduct research or other activities with regard to the indoor environment. The Tennessee Valley Authority (TVA) provides public information and education regarding indoor air quality, technical assistance to agencies and organizations, and research projects. The General Services Administration (GSA) is involved in indoor air quality as it relates to its management of federal buildings. The National

Institute of Standards and Technology has developed an indoor air quality research model, and the National Institute of Building Science works with the building industry to integrate measures that promote indoor air quality into building practices.

EPA's report to Congress

Among the most serious health threats to be recognized was that of radon. In October 1989 President Reagan signed into law the Indoor Radon Abatement Act as Title III of the Toxic Substances Control Act. The goal of this act is to bring indoor radon levels down to that of ambient outdoor air. Among its mandates, the law required the establishment of building codes and radon abatement techniques and demanded that all federal buildings undergo radon testing, with results submitted to EPA and a report to Congress by October 1990. In addition, under the law EPA must develop and implement activities designed to assist state radon programs and submit an annual report to Congress identifying state projects. It made available $10 million for each of the following three fiscal years to be used as seed money for the development of state radon programs. EPA was also charged with compiling a comprehensive study and survey on radon in the nation's schools.

In 1989 EPA's Report to Congress on Indoor Air Quality was released. Noting that "indoor air pollution represents a major portion of the public's exposure to air pollution and may pose serious acute and chronic health risks," the agency recommended a six-point expanded effort to "characterize and mitigate this exposure." Among the suggestions were:

EPA's recommendations to Congress

1. Research to characterize exposure and health effects of chemical contaminants;
2. Research to characterize and develop mitigation strategies for biological contaminants;
3. Research to identify and characterize indoor air pollution sources and to evaluate appropriate mitigation strategies;
4. A program joining with private sector groups to develop and promote guidelines covering ventilation, building design, operation and maintenance practices for ensuring indoor air quality;
5. A program of technical assistance and information dissemination, including an indoor air quality clearinghouse, to inform the public about risks and mitigation strategies and to assist state and local governments and the private sector in solving indoor air quality problems;
6. An effort to characterize the nature and pervasiveness of the health impacts associated with indoor environments in commercial and public buildings, and develop and promote recommended guidelines for diagnosing and controlling problems.

The report included a recommendation for a five-year indoor air pollution research plan costing $99.15 million.

That same year a Massachusetts Special Legislative Commission on Indoor Air Pollution estimated that indoor air pollutants account for more than half of all illnesses and cost the country more than $100 million a year in medical bills and sick leave.

Mounting evidence of the seriousness of this problem and the insufficiency of federal efforts in research and mitigation have led to the proposition of the Indoor Air Quality Act of 1990, which at the time of writing, was still before the Senate and the House of Representatives.

Such legislation would authorize funds (estimated at close to $50 million each year through 1995) for items such as:

- continued research (with particular emphasis on technology and management practices, childcare facilities, exposure assessments and ventilation standards);
- publication of indoor air contaminant health advisories;
- development of a national indoor air quality response plan;
- establishment of an EPA Office of Indoor Air Quality;
- establishment of an indoor air quality clearinghouse of information;
- funding of state and local indoor air quality programs;
- establishment of a Council on Indoor Air Quality (including several federal agencies such as the EPA, OSHA, NIOSH, DHHS, HUD, DOE, CPSC, and GSA);
- development of a building assessment demonstrations program.

Although the suggested funding is small in comparison with that devoted to combatting outdoor pollution, it indicates a great improvement over the less than $10 million spent in the entire decade of the seventies.

Nevertheless, in its August 1989 issue, *Indoor Pollution Law Report*, a monthly newsletter reporting on this growing area of law, noted that Barbara J. Katz, of the Consumer Federation of America cited research by the American Council for an Energy Efficient Economy that indicated that the federal government spends roughly $6 per person per year on research related to outdoor air in contrast to six cents per person each year on indoor air research.

She made these remarks at a House Subcommittee hearing on the proposed legislation. This bill has elicited a wide range of responses as evidenced by the long debate over its passage. The EPA remains opposed to the bill on the grounds that it is inappropriate since the agency has already been mandated to act on the issue under the Superfund Amendments. In addition, although the EPA is in agreement with the bill's call for increased research, the agency believes that the long list of priorities indicated in the bill are too prescriptive and inflexible, restricting the agency's ability to assign priorities as needs arise. On another level, the EPA looks askance at the bill's technology demonstration plan, which calls for EPA and GSA to develop and implement a program of reduced indoor pollution and demonstrate their methods in new federal buildings.

Other criticisms of the bill score it for creating overlaps among the many institutions involved; placing too much emphasis on mitigation strategies and not enough on practical solutions such as ventilation; and paying little or no attention to biological pollutants.

Involvement at the state and local levels

Along with the federal government, the states are finding themselves increasingly involved in indoor air quality issues. Efforts are usually divided among a variety of agencies and programs, each focusing on a particular pollutant (e.g. radon, asbestos, pesticides) or setting (schools, hospitals). State health agencies in most states are active in this area, as are departments of energy, building codes and standards, agriculture and education. Still, many of the programs at the state level are still ill-defined or lacking proper funding. A survey of state health agencies in 1988 found that of 32 such agencies reporting that they assign staff to indoor air quality issues, only five reported they have general IAQ programs, separate from programs on specific issues such as radon.

State involvement is mainly in the form of disseminating information about indoor air pollution to the public, and responding to public inquiries. Some states also provide technical assistance or training to those involved with IAQ activities, monitor indoor air and investigate reported problems, and do some basic research.

Awareness of the importance of indoor air quality can be easily seen as the number of court cases involving indoor air pollution grows with each passing year. Builders, architects, ventilation systems contractors, home inspectors, manufacturers of building products, ventilation companies, building owners and managers, exterminators, home sellers, large industry and small business owners are all finding themselves worrying about their liability with regard to indoor pollutants. Home and land buyers, workers, and consumers are finding their way to court when perceived exposure to indoor pollutants is followed by unusual illnesses or symptoms. Insurance companies are being warned to expect a wide range of indoor pollution-related claims, especially in the area of radon exposure.

Indoor pollution-related lawsuits on the rise

A recent verdict, in one of the first lawsuits filed involving the issue of radon contamination, awarded the buyers a full refund of their down-payment after the home was found to contain an unsafe radon level. In September 1989 the purchasers, Williams Living Trust, agreed to purchase property in Roaring Brook Township, Pa.. The sales agreement contained a Radon Disclosure Addendum that provided that the purchasers would arrange and pay for a radon test of the home within 10 days from the date of the agreement. If the test verified radon at a level exceeding 4pCi/l the sellers were required to submit a corrective proposal to the purchasers, who had five days either to accept the proposal or back out of the agreement and receive a return of their deposit.

After tests revealed radon levels slightly higher

than 4pCi/l the purchasers notified the sellers, who did not comply with the agreement but offered to pay half the total cost of the remedial action necessary to correct the problem. The purchasers declined this offer and demanded a return of their deposit. The sellers refused and the case went to the Court of Common Pleas of Lackawanna County, Pa. as Williams Living Trust v. Calpin.

There is, of course, little legal precedent in this area, making it difficult to keep similar cases out of court. Sellers presented with elevated radon levels are typically unwilling to accept the idea that there is anything wrong with their house. Buyers are justifiably concerned about the health risks. On the plus side, however, is the fact that effective radon mitigation is not very costly. In known "hot" areas, real estate contracts are beginning to sprout radon disclosure addendums or contingency clauses that allow the purchaser to back out of the deal if tests reveal radon in excess of the EPA's accepted level.

If such a level is found and the buyers walk away from the property, the seller is then required to notify all subsequent purchasers of the information, whether or not a radon disclosure addendum is part of the new contract. The intentional witholding of such information (even if the purchaser does not ask any questions) would be considered tantamount to the failure to disclose a known defect if the case ever arose in the future. In addition, realtors or other parties involved in the sale who know of the elevated levels may also be liable for failing to disclose the information if they are involved with subsequent buyers of the house.

Evaluate potential problems first

Builders and developers are also urged to proceed with caution before buying land or building in areas known to have a radon problem. They are advised that the best time to evaluate potential problems is before they buy or even build property. Lawyers are advising such clients that elevated radon can result in a basis for liability against builders, developers or landlords. In the August 1989 issue of *Indoor Pollution Law Report*, an article advises that: "If the builder knows that radon is a problem in a certain area but does not disclose it, the purchaser may sue for breach of warranty, negligence or fraud, or for unfair or deceptive trade practices."

In order to avoid this scenario, builders and developers are urged to test for soil radon levels around the property and map the levels, indicating any "hot spots" of extremely high levels, where building should be avoided or stringent measures taken to mitigate the entry of radon into buildings. Unfortunately, knowledge of soil testing and analysis of the results are still in their infancy, putting their reliability into question.

The presence of other indoor pollutants has also led to the courtroom, including a recent Mississippi case in which a jury awarded $250,000 to a family that claimed to suffer severe damage to their immune systems as well as multiple chemical sensitivities due to an exterminator's negligence in applying the pesticide Aldrin to their home to eradicate termites.

In another pesticide case, an Arizona jury awarded $760,000 to an elderly couple when the exterminating firm Truly Nolen of America, Inc. illegally sprayed chlordane into the crawl space and around the perimeter of their house, resulting in acute poisoning and forcing them to desert their home. It took more than a year before clean-up efforts brought indoor air levels below the 5 micron per cubic meter ceiling recommended by the National Academy of Science.

In mid-1989 a Georgia Superior Court jury awarded $270,000 in compensatory damages and $130,000 in punitive damages to the Radtke family, when their home was rendered uninhabitable after Arrow Exterminators sprayed a pesticide containing chlordane and heptachlor. The family testified they had suffered headaches, eye disturbances and upper respiratory ailments after the pesticide treatment to their new home and one family member claimed to suffer auto-immune stomach and liver disease.

In the case of David Pinkerton et al. v. Georgia-Pacific Corp. and Temple Industries, the family was awarded $16 million in punitive damages plus another $203,000 in compensatory damages in Clay Country, Mo., when the high level of formaldehyde outgassing from particleboard in their newly constructed home caused them to suffer severe health problems, permanent damage to various organ systems and to their immune systems, and significantly increased risk of cancer. Formaldehyde levels in the home (measured repeatedly at between 3 and 10 ppm) greatly exceeded the NIOSH 0.1 ppm limit for eight-hour occupational exposure, and the family was forced to leave the home they had been living in for more than a year. The suit was based on theories of strict liability, failure to warn and implied warranty of merchantability. The manufacturers of the materials were found to have chosen to continue sales of the particleboard product with a warning attached indicating it was merely an "irritant," despite a growing body of research on the formidable health risks of formaldehyde. This case appears to have marked the first time punitive damages were awarded because of formaldehyde exposure.

Asbestos litigation has been among the biggest sources of cases to come to courts around the country. Other incidences of indoor pollution as the basis for legal action include outgassing of toxic chemicals from carpet backing, faulty ventilation systems leaking carbon monoxide, carbon dioxide and other pollutants into the indoor environment, and radiation emitted from video display terminals. In additional to these, of course, are the many claims of Sick Building Syndrome without the appearance of an individual pollutant as the cause.

Clearly, the quality of the air we breath indoors has become of the greatest concern, both in America and around the world, as attested to by the proliferation of international indoor air qual-

The public deserves to be informed

ity conferences. At a 1989 meeting of the North Atlantic Treaty Organization (NATO) Committee on the Challenges of Modern Society, Robert Axelrad, Director of the EPA's Indoor Air Division, stressed that: "The relative risks from indoor air pollution are high compared to many other environmental problems and the public deserves to be informed of these risks and the steps that can be taken to reduce these risks. Many actions to reduce risks will remain nonregulatory, and therefore in the sphere of informed choice."

As this book has tried to say: The health consequences of the quality of the air in your home are real. The ability to breathe clean is largely up to you.

8 The Last Word

I have a good friend who regularly vacuums, makes daily use of the dusting spray and frequent use of the spray polish, routinely cleans her carpets and upholstered furniture with the latest consumer products, keeps a spotless oven with the help of spray-on oven cleaner, and generally works hard to give her family the advantages of cleanliness. In fact, what she and her family may receive is a substantial dose of propellents, other assorted hydrocarbons and organic compounds, and the daily resuspension of airborne fine particulates.

I, on the other hand, come from a very clean and tidy family, where my personal tendency toward clutter and generally lackadaisical attitude toward housekeeping was not well accepted. For years I felt guilty about my inability to dust, sweep, and vacuum my own apartment every day. Imagine my delight when I discovered that all that dusting and vigorous cleaning really wasn't very healthful after all—I was vindicated!

Whither housekeeping?

Obviously, we have to clean our homes, but when even cleaning itself proves to be a source of pollution it may be hard to know where to turn. Whether you're a compulsive housekeeper or an intermittent one, given all that you now know about the importance of indoor air quality, and the health threat of indoor pollution, how can you go on living in your home or apartment? The answer is not, "I can't," followed by a frenzied call to the real-estate agent and a one-way ticket to Tahiti. The real, logical answer is that you need to assess carefully the extent of potential pollution in your own environment and make the proper adjustments to minimize their impact on your family's health.

Ventilation and pollution sources

One critical point is that your home's level of pollution is heavily dependent on two factors: Its level of ventilation in the form of fresh, outdoor air, and the number and strength of the pollution sources it harbors. Energy-efficient buildings are not inherently plagued by indoor air quality problems, but they are more likely to develop problems if they contain many strong sources of contaminants, such as gas stoves and ovens, unvented kerosene space heaters, urea-formaldehyde foam insulation, tobacco smokers, lots of plywood and particleboard furnishings, and the constant emanation of radon gas from the soil below.

Indoor air quality can be be controlled in two basic ways. Pollution sources may be isolated, disposed of, or altered to prevent contaminants from entering your personal environment, or the existing contaminants can be removed from the air, either through increased ventilation with clean air or use of effective air cleaners. Any of these practical strategies involves some effort and perhaps even financial outlay on your part and so should be carefully considered.

In the end you need to do as much as you feel comfortable with. One family may be able to retrofit their home with heat exchangers, trade their gas appliances for electric ones, and purchase a battery of high-efficiency air cleaners. Another might feel that banning tobacco smoking in the home and installing an outside-vented range hood over the gas stove is acceptable and efficient. They have both improved the quality of their indoor air and taken measures to safeguard their health. Perhaps it would help you to examine your needs through a three-step process. First, assess any recurrent health problems from which you or members of your family suffer, and any special needs they may have. Is there a tendency toward:

Your symptoms

- itchy, irritated, or burning eyes?
- constant runny noses, lingering colds?
- attacks of coughing or sneezing?
- difficulty in breathing?
- chronic headache?
- itchy skin or periodic rashes?
- unaccountable lethargy or irritability?

Some of these may be allergic or toxic responses to contaminants in your indoor air, although symptoms may also be linked with other causes and, taken alone, are not reliable indicators of the presence of contaminants. Now examine these symptoms more closely:

Detecting a pattern

- Do they abate when you leave your home for any length of time?

- Do they occur or worsen at regular times of the day? When?

- Do they occur or worsen in specific locations? Where?

- Has a consultation with your doctor resulted in no conclusive diagnosis?

- Do other family members or guests in your home or apartment share these symptoms and their patterns of occurrence?

If you answer yes to one or more of these questions and there is a relationship between the symptoms and certain rooms in the house, the possibility is greater that there is an air quality problem. As for special needs, infants, young children, the elderly, and those confined to the home for more than 12 hours per day are at the highest risk of suffering adverse health effects from indoor air contaminants. The presence of any of these high-risk household members signals a need for particular attention to indoor air quality.

In addition to noticing a pattern of adverse health effects, you may be able to recognize deteriorating indoor air quality by the presence of strange or foul lingering odors, or dampness during the winter months. Detecting odors is not the best indicator of the presence of contaminants, however, because the ability to smell a specific odor tends to diminish over time. In addition, strong odors will mask any weaker ones that are present, and each person has a different level of odor sensitivity. Regardless of whether you suspect that something in the air is producing demonstrable health symptoms, you should examine your home for its pollution potential by answering the next set of questions:

Everyday habits

- Do you live in a mobile home or a newly constructed home with a great deal of plywood or particleboard?
- Have you recently added insulation to your house?
- Have you replaced old windows? Did you replace them with double- or triple-glazed models?
- Do you have a gas or oil furnace or hot water system?
- Do you have an unvented gas stove and oven?
- Do you ever use your gas oven as a heat source?
- Do you use unvented space heaters?
- Do you supplement your heating system by burning wood or coal in a fireplace or stove?
- Do you use a humidifier to replace moisture in the air?
- Do you live in an area that is now, or was ever, mined?

- Is your home on a landfill site?
- Do you live in an area of granitic rock or an area known to have a high radon content? Do you use well-water?
- Do you engage in hobbies such as carpentry, metal work, or model building? Are paints, solvents, or other supplies stored indoors?
- Do you use a wide array of housecleaning products, especially spray products?
- Does anyone in your household smoke cigarettes, cigars, or a pipe?
- Does your house have an attached garage?
- Have pesticides, fungicides, or wood preservatives recently been used around your home or in its building materials?
- Have you noticed frayed insulation around basement pipes or seen a fibrous dust on household objects?

Plan changes

If you answer yes to two or more of these questions, your indoor air quality is probably affected to some degree. If, in addition, you know that you have a tight, energy-efficient house, your indoor air quality and the health of you and your family would almost certainly benefit by taking steps to lower the overall pollution level and the concentration of your own key contaminants. These can be identified with the help of the "Walk Through the House" checklist or by professional air testing. In the accompanying table, we take one more "walk through the house," to pinpoint our troublespots and plan the changes that can make a difference in the way we live.

A final "walk"

The Final Wrap-Up: Easy Ways to Improve Your Indoor Air Quality

Room	Pollution source	Steps you can take to lower pollution output
Kitchen	Appliances	Install electric range and oven. If you have a microwave oven, do not slam door shut, fling door open, or use door as a shelf. When in doubt about proper safety, have oven professionally serviced and tested. If you have a gas stove and oven, have it vented to the outdoors and always use hood vent when you are cooking. If you cannot install an outside vent, open kitchen window when stove or oven is in use. To further reduce pollution, place an exhaust fan in the window to draw out emissions. Make sure gas stove and oven burners are properly

	adjusted (flame should burn blue, not yellow).
	If you purchase a gas stove or oven, buy only pilotless appliances.
	Do not use gas clothes dryers unless they are vented to the outside. Make sure exhaust vents are not near windows or air-intake vents, to prevent re-infiltration.
	Never use gas oven as a supplemental heat source.
Cleaning products	Avoid aerosol spray cans whenever possible.
	Substitute low-polluting cleaning products and techniques whenever possible. (See Product Substitution Table in chapter 4 for suggestions.)
Miscellaneous	Keep only a bare minimum of paper bags and other paper products.
	Avoid aluminum cookware and utensils.
	Install and use an efficient air

		cleaner (such as HEPA air filter, electrostatic precipitator, or ion generator) to minimize pollutants.
Living room	Fireplace	Install an air duct at bottom rear of fireplace to allow the fire to draw in outside air for combustion.
		Keep burning wood well inside the boundaries of the fireplace.
		Avoid burning synthetic fireplace logs, coated stock paper, color-printed paper, and newspaper.
		Glass fireplace doors may cut down on particulate emission.
	Wood stove	Avoid burning coal.
		Have wood stoves, flues, and chimneys properly installed, allowing ample room for smoke to dissipate far from windows and air vents.
		When shopping for a wood stove, look for one with a secondary combustion chamber and/or catalytic

converter that reduces the emission of hydrocarbons. Some stoves also come with a direct air feed from the outside.

To avoid back drafts of smoke when opening stove door, first open the damper to increase the draft up the flue and then open the door.

Periodically check gaskets and joints to make sure they are tight.

Home furnishings

Try to buy furniture made of solid wood rather than veneer-covered particleboard.

Although this is not very scientific, take a good sniff of carpets and carpet pads before purchasing (especially for wall-to-wall installation). If you have a physical reaction, or smell a strong, pungent formaldehyde odor, choose another product.

In general, hardwood floors and small area rugs hold less dust than wall-to-wall carpeting and are less likely to be strong sources of

		formaldehyde and mold spores.
		Have carpets and rugs professionally steam-cleaned.
		Avoid home carpet-cleaning sprays.
	Tobacco smoke	Prohibit tobacco smoking in this and other rooms, or designate one well-ventilated room for smoking.
		Use an efficient air cleaner. (See chapter 4 for the range of types.)
Bedrooms and nursery	Space heaters	Use only outside-vented gas or kerosene space heaters. Use electric heaters, oil-filled electric radiators, or electric baseboard heaters to supplement home heating system.
	Humidifiers	Clean all humidifiers daily with a strong solution of vinegar and hot water.
		Substitute ultrasonic humidifiers for cool-mist models.

		If you do use a cool-mist humidifier for occasional use, start it with very hot water, which harbors fewer microorganisms.
		Never allow water to stand in the humidifier for long periods and never reuse water from previous use.
	Air conditioners	Service and repair air conditioners when needed. Clean coils and check them occasionally for visible microbial growth.
	Miscellaneous	Use an air cleaner where needed to keep bedroom and nursery air particularly clean. This can be especially important for allergy or asthma sufferers. (See chapter 4 for specific types of air cleaners.)
Bathroom	Fixtures	Keep toilets, tubs, and sinks clean and mold-free.
		Dry off tile walls (especially round tub and shower) to discourage microbial growth on damp surfaces and clean them often.

		Clean any mold or mildew with a strong solution of vinegar and hot water.
		Use oxygen-bleach abrasive cleansers instead of chlorine-bleach ones.
	Personal Products	Avoid aerosol cans whenever possible.
		Discard any half-used products you no longer need.
		Avoid hair spray or any personal product that contains methylene chloride.
	Miscellaneous	Ventilate bathrooms especially well. A simple exhaust fan does wonders.
		Avoid the use of chemical air fresheners.
		Use drain openers sparingly and carefully and never use in conjunction with any other cleaning product.
All rooms	House cleaning	Replace regular portable vacuum cleaner with central vacuum system.

		Avoid spray products for dusting, polishing furniture, or spot-cleaning carpets. Substitute low-polluting cleansing products for high-polluting ones. (See chapter 4 for suggestions.)
		Brush and groom hairy pets often—outdoors.
		Use silicone-treated dustcloths to hold the dust rather than resuspend it in the air.
Basement	Furnace	Replace combustion systems with electrical systems.
		If you have a gas- or oil-burning system, check burners for proper adjustment and check flues and other pipes for cracks or leaks. Repair wherever needed.
	Air ducts	Check air ducts for corrosion or any loose, fibrous material, and repair where needed.
		Install an efficient filter system in air ducts to eliminate particulates.

		Install ventilating heat exchangers.
	Pipe insulation	Check for shredding or deteriorating pipe insulation and repair or replace. (If you believe you have asbestos pipe insulation, get a professional to do the work.)
		Seal fibrous pipe insulation with a penetrant composed of vinyl acrylic polymers and inorganic material.
Garage		Never idle motor vehicles in an attached garage.
		Separate garage and house.
Building superstructure		Seal all cracks in basement walls and concrete slabs with polymeric caulks to prevent infiltration of radon gas from soil.
		Ventilate crawl spaces.
		Cover basement walls and concrete slabs with epoxy paint, polymeric sealant, or polyethylene or polyamide film vapor barrier to prevent

introduction of radon from surrounding soil.

In an area known to have high-radium-content soil, excavate local soil beneath foundation and replace with low-radium-containing soil.

Replace plywood, particleboard, and fiberboard building elements with solid wood elements where possible.

Do not build in areas of uranium or phosphate mining or where tailings have been used for landfill.

Avoid building materials with high radon content (where possible).

Use HEPA filtration, electrostatic precipitation, or ion generation as needed to reduce radon progeny.

Cover particleboard, plywood, and fiberboard with shellac, varnish, polymeric coating, or other diffusion barriers.

		Avoid use of urea-formaldehyde foam insulation.
		Remove need for insecticide by constructing termite-proof foundations.
Miscellaneous	Hobbies	Do all wood sanding in a well-ventilated room separate from the rest of the house.
		Use solvents in a well-ventilated room, separate from the rest of the house and occupants.
		Seal partially used paints, solvents, strippers, and other chemical compounds tightly and store in a well-ventilated area away from air ducts.
		Discard products you will not use again.
	Pesticides	Eliminate need for pesticides by constructing termite-resistant foundation.
		Replace indoor insecticides with bug traps.

	To rid houseplants of pests, wash plants often in warm soapy water and rinse with clear water.
	Especially avoid aerosol spray insecticides.
Lighting fixtures	Replace old fluorescent fixtures with new, preferably incandescent lighting.

Additional Sources of Information

Federal Agencies

U.S. Environmental Protection Agency (EPA)
Public Information Center
401 M. Street SW
Washington, DC 20460
202–382–2080 and 2090
The EPA has published a great number of booklets aimed at acquainting the public with the problem of indoor pollution and some of the solutions. Among them are: *Survey of Indoor Air Quality Diagnostic and Mitigation Firms* (listing some 1,200 sources) and *Directory of State Indoor Air Contacts*. These are both extensive works and highly useful. In addition, EPA produces many simple brochures on aspects of indoor air quality.

U.S. Consumer Product Safety Commission (CPSC)
Washington, DC 20207
800–638–CPSC or 800–638–2772

National Pesticides Telecommunications Network
800–858–PEST or 800–858–7378

Toxic Substances Control Act Assistance Information Service
202–554–1404

Safe Drinking Water Hotline
202–382–5533

Resource Conservation and Recovery Act/Superfund Hotline
800–424–9346
In Washington 202–382–3000

U.S. Department of Health and Human Sevices NIOSH Public Information
800–35–NIOSH or 800–356–4674

Indoor Air Quality Assistance
513–841–4382

State Radon Hotlines
Alabama 205–261–5313
Alaska 907–465–3019
Arizona 602–255–4845
Arkansas 501–661–2301
California 415–540–2134
Colorado 303–331–8480
Connecticut 203–566–2275
Delaware 800–554–4636
D.C. 202–783–3183
Florida 904–488–1344
Georgia 404–894–6644
Hawaii 808–548–4383
Idaho 208–334–5927
Illinois 217–786–6398
Indiana 317–633–0153 or 800–272–9723
Iowa 800–383–5992
Kansas 913–296–1560
Kentucky 502–564–3700
Louisiana 504–925–4518
Maine 207–289–4518
Maryland 800–872–3666
Massachusetts 617–727–6214
Michigan 517–335–8190
Minnesota 800–652–9747
Mississippi 601–354–6657
Missouri 314–751–6983 or 800–669–7236
Montana 406–444–3671
Nebraska 402–471–2168
Nevada 702–885–5394

New Hampshire 603–271–4674
New Jersey 800–648–0394
New Mexico 505–827–2984
New York 800–458–1158
North Carolina 919–733–4283
North Dakota 701–224–2348
Ohio 800–523–4439
Oklahoma 405–271–5221
Oregon 503–229–5797
Pennsylvania 800–23–RADON
Rhode Island 401–277–2438
South Carolina 803–734–4700
South Dakota 605–773–3153
Tennessee 615–741–3931
Texas 512–835–7000
Utah 801–538–6734
Vermont 802–828–2886
Virginia 800–468–0138
Washington 206–753–5962 or 800–323–9727
West Virginia 304–348–3526 or 800–468–0138
Wisconsin 608–273–6421
Wyoming 307–777–7956

GLOSSARY

Ach Abbreviation for air changes per hour, a unit of air exchange rate.

Adsorption Removal of gaseous pollutants from the air by means of their retention on the surface of a solid material filter.

Aerosols Solid or liquid particles small enough to remain suspended in the air for a period of time, and settle out slowly under the force of gravity.

Air exchange Movement of air into and out of a structure by means of infiltration, exfiltration, natural ventilation, and mechanical ventilation.

Air exchange rate Amount of air flowing into and out of a structure during a given time period.

Aldehydes A series of organic chemicals containing carbon, hydrogen, oxygen groups (CHO groups), and having strong odors.

Aliphatic hydrocarbon An open-chain hydrocarbon, such as propane, methane, and butane.

Allergen A substance that produces adverse health effects in only a portion of the population.

Alpha particle A positively charged particle (a helium nucleus) emitted in the decay of a radioactive element such as radon.

Alveoli	The tiny air sacs in the lung where oxygen is transferred to the blood and carbon dioxide is taken in exchange.
Ambient air	According to the Environmental Protection Agency, "that portion of the air, external to buildings, to which the general public has access."
Aromatic hydrocarbons	Organic chemicals characterized by the presence of the benzene-ring molecular structure.
ASHRAE	American Society of Heating, Refrigeration, and Air Conditioning Engineers.
Asphyxia	Suffocation.
Breach of contract	Failure to fulfill a contractual agreement.
Breach of warranty	Failure of a product (or process) to fulfill express or implied promises or representations.
Bronchial tubes	Branches of subdivision of the windpipe (trachea).
Bronchiole	A minute, thin-walled branch of a bronchial tube.
cfm	Abbreviation for cubic feet per minute, a unit of air flow rate.
Ci	Abbreviation for Curie, a unit of radioactivity equal to 37 billion disintegrations per second.

Carcinogenic Producing cancer.

Concentration Amount of a contaminant in a given quantity of air.

Contaminant Substance in the air that is generally not present or that is present in greater-than-normal quantities.

Diffusion Spontaneous movement of gas molecules or particles throughout the air from areas of high concentration to areas of low concentration.

Dose Quantity of a substance absorbed in a part of the body or in an individual.

Efficiency Effectiveness of a device in removing particles from the air, often expressed as the percentage of particles originally in the air that have been removed.

Electrostatic precipitation Removal of particles from the air by imparting a charge to them and then attracting them to an oppositely charged collection material.

Embryotoxic Poisonous to a developing fetus (see fetotoxic).

Emission rate Amount of pollutant released by a source in a specific amount of time.

Encapsulation Completely covering or coating an object, such as a pollution source, with a film or coating to prevent the release of contaminants.

Envelope	The exterior walls, windows, doors, and roof of a building that enclose the interior space.
EPA	Environmental Protection Agency.
Exfiltration	Movement of air out of a building through cracks and interstices in the building envelope.
Fetotoxic	Poisonous to a developing fetus (see embryotoxic).
Filtration	Removal of particles from air or water by means of passing the air through a filter that screens out the particles.
Fungus	A plant, lacking chlorophyll, that lives off other living organisms.
Half-life	The time it takes for half of the atoms in a given quantity of a radioactive isotope to decay.
Heat exchanger	A device for recapturing a portion of the heat in one airstream and transferring it to another airstream without physical contact between the two airstreams.
HEPA	Abbreviation for high efficiency particulate air filter, an extended surface filter with a high rate of particle removal.
Hemolytic	Producing separation of hemoglobin from the red blood cells.
Hydrocarbon	Compounds containing only hydrogen and car-

bon, such as methane, the major component of natural gas.

Infectious agents Viruses, bacteria, and microorganisms that can cause human disease.

Infiltration Movement of air into a building through cracks and interstices in the building envelope.

Inhalable particles Particles (usually smaller than 15 micrometers) that can bypass filtrations by the nose and be deposited along the respiratory tract.

Ion generator A device, often used for air cleaning, which produces and disseminates highly charged particles into the air.

Ionization Removal of an electron from an atom by imparting energy and thus creating a charged (negative or positive) particle.

Mechanical ventilation Forced movement of air by fans into and out of a building.

Microgram A unit of mass equal to one millionth of a gram.

Micrometer A unit of length equal to one millionth of a meter.

Micron A micrometer.

Millimeter A unit of length equal to one thousandth of a meter.

Molecule Generally, the smallest particle of a substance that has the same chemical properties as the larger mass and is composed of one or more atoms.

Mutagenic Inducing genetic mutation.

Natural ventilation Movement of air into and out of a building through intentional openings, such as windows and vents.

Negligence Failure to exercise proper care, resulting in responsibility for injury or damage to a person or thing.

Neuritis Inflammation of a nerve.

Nanogram A unit of mass equal to one billionth of a gram.

NIOSH Abbreviation of National Institute of Occupational Safety and Health.

Organic chemicals Chemical compounds that contain carbon with the usual exception of carbon dioxide (CO_2).

OSHA Abbreviation for Occupational Safety and Health Administration.

Outgassing Emission of gases during curing, aging, and degradation of a substance.

Particulate Small particle that remains suspended in the air, or an adjective indicating that a material is composed of particles.

Pollutant A substance normally not found in air or water or one that is present in higher-than-normal quantities that can produce adverse health effects.

ppm Abbreviation for parts per million, a unit of concentration.

Prefilter An initial filter that removes the largest particles in an air cleaner.

Radiation Electromagnetic waves; emission of electromagnetic waves; or energy transfer by particles (such as alpha particles).

Radon progeny The radioactive elements resulting from the decay of radon.

Respirable particles Particles that penetrate to the lungs when inhaled; usually particles smaller than five micrometers.

Smoke Mixture of gases and small particles generated by incomplete combustion.

Source Object or process that releases pollutants into the air or water.

Strict liability Responsibility for injury or damage in spite of the fact that the terms of the contract or warranty have been fulfilled.

Suspended particles Particles (usually smaller that 100 micrometers) that are small enough to remain in the air and settle out slowly under the force of gravity.

Synergy The combined action or effects that are generally greater than the sum of their individual effects.

Teratogenic Tending to produce anomalies of formation in an embryo.

Toxic Able to produce adverse health effects after physical contact, ingestion, or inhalation.

Vapor Gaseous phase of a substance that is predominantly a liquid or solid at room temperature.

Ventilation Controlled movement of air into and out of a building.

Volatile organic chemical A carbon-containing compound that tends to evaporate rapidly under normal atmospheric conditions.

WL Abbreviation for working level, a measure of radon progeny concentration; any combination of radon progeny in one liter of air whose radioactive decay will result in the release of 130,000 million electron volts (MeV) of alpha-particle energy.

Notes

Chapter 1

Indoor Pollution: The Unsuspected Threat in Your Home.

1. Szalai, A. ed., *The Use of Time: Daily Activities of Urban and Suburban Populations in Twelve Countries* (The Hague: Mouton & Co., 1972), cited in National Research Council, *Indoor Pollutants* (Washington, D.C.: National Academy Press, 1981), p. V 3

2. Oak Ridge National Laboratory, "Status Report on Indoor Air Quality Monitoring Study in 40 Homes" (Washington, D.C.: Consumer Product Safety Commission, 1984)

Chapter 2

A Walk Through the House

1. National Research Council, *Indoor Pollutants* (Washington, D.C.: National Academy Press, 1981), chapter IV

2. Spengler, John D. et al., "Nitrogen Dioxide Inside and Outside 137 Homes and Implications for Ambient Air Quality Standards and Health Effects Research," *Environmental Science & Technology* (Vol. 17, No. 3, 1983), pp. 164–168

3. Spengler, John D. et al., "Sulfur Dioxide and Nitrogen Dioxide Levels Inside and Outside Homes and the Implications on Health Effects

Research," *Environmental Science & Technology* (Vol. 13, No. 10, Oct. 1979), pp. 1276–1280

4. Traynor, G.W. et al., Indoor Air Quality: Gas Stove Emissions (Berkeley, Cal.: Lawrence Berkeley Laboratory Report 1979)

5. Vedal, Sverre, "Epidemiological Studies of Childhood Illness and Pulmonary Function Association with Gas Stove Use," *Indoor Air and Human Health*, Gammage R.B. and Kaye S.V., eds. (Chelsea, Mich.: Lewis Publishers, 1985), p. 304

6. National Research Council, *Indoor Pollutants*

7. Ibid.

8. Meyer, Beat, *Indoor Air Quality* (Reading, Mass.; Addison-Wesley Publishing Co., Inc., 1983)

9. "The New Wave in Microwave Ovens," (*Consumer Reports*, Nov. 1985), p. 647

10. Calle, E.F. and Zeighami, E.A., "Health Risk Assessment of Residential Wood Combustion," *Indoor Air Quality*, Walsh, P.J., Dudney, Charles S., and Copenhaver, E.D. (Boca Raton, Fla.: CRC Press, 1984), pp. 39–53

11. Kreiss, K. et al., "Respiratory Irritation Due to Carpet Shampoo: Two Outbreaks," *Environment International*, (Vol. 8, Nos. 1–6, 1982), pp. 337–341

12. "Kerosene Heaters," (*Consumer Reports*, Oct. 1982)

13. Traynor, G.W., et al., Indoor Air Pollution Transport Due to Unvented Kerosene-Fired Space Heaters, (Berkeley, Cal.: Lawrence Berkeley Laboratory Report, 1984)

14. Turiel, Isaac, *Indoor Air Quality and Human Health* (Stanford, Cal.: Stanford University Press, 1985)

15. "Ultrasonic Humidifiers," *Consumer Reports* (Nov. 1985), p. 682

16. Shabecoff, Philip, "Radioactive Gas in Soil Raises Concern in Three-State Area," *New York Times*, (May 19, 1985)

17. National Research Council, *Indoor Pollutants*, pp. VII–67–76

18. Turiel, Isaac, *Indoor Air and Human Health*, p. 70

19. Spengler, John D., and Cohen, Martin A.: "Emissions from Indoor Combustion Sources," *Indoor Air and Human Health*, Gammage, R.B. and Kaye, S.V. eds., pp. 271–273

20. National Research Council, *Indoor Pollutants*, pp. VII–64–67

21. Hirayama, T., "Nonsmoking Wives of Heavy Smokers have a Higher Risk of Lung Cancer: A Study from Japan," *British Medical Journal* (Vol.

282, 1981) pp. 183–185 cited in National Research Council: *Indoor Pollutants*.

Chapter 3

A Walk Through the Office: A Stop at the School

1. Kreiss, Kathleen, et al., "Respiratory Irritation Due to Carpet Shampoo: Two Outbreaks," *Environment International* (Vol. 8 No. 1–6), pp. 337–341

2. Kreiss, Kathleen and Hodgson, Michael, "Building-Associated Epidemics," *Indoor Air Quality*, pp. 87–106

3. Kreiss, Kathleen and Hodgson, Michael, "Building-Associated Epidemics," p. 99

4. Ibid.

5. Miksch, R.R., et al., "Trace Organic Chemical Contaminants in Office Spaces," *Environment International*, pp. 129–137

6. Ibid.

7. Meyer, Beat, *Indoor Air Quality*, pp. 70–71

8. Wadden, R.A., and Scheff, P.A., *Indoor Air Pollution*, pp. 70–71

9. Brody, Jane, "Surprising Health Impact Discovered for Light," *New York Times* (Nov. 13, 1984), pp. C1, C3

10. Ibid.

11. Ibid.

12. Kreiss, Kathleen and Hodgson, Michael, "Building-Associated Epidemics," p. 97

13. Kreiss, Kathleen et al., "Respiratory Irritation Due to Carpet Shampoo," pp. 338–339

Chapter 4 **What You Can Do to Lower Indoor Pollution**

1. "Heat-Recovery Ventilators," *Consumer Reports* (Oct. 1985) pp. 596–599

2. Hasbrouck, Sherman and Breece, Linda, "Removing Radon from Water Using Granular Activated Carbon Adsorption," (Orono, Me.: University of Maine at Orono, 1983)

3. "A Test of Small Air Cleaners," *New Shelter* (July/August 1982)

4. Offerman, F.J. et al., "Control of Respirable Particulates and Radon Progeny with Portable Air Cleaners," (Berkeley, Cal.: Lawrence Berkeley Laboratory Report, 1983)

Chapter 5 **Should the Government Regulate Indoor Air Quality?**

1. Blagg v. Fred Hunt Corp., Inc., *South Western Reporter*, 2nd Series, 612: 321–322 (1981)

2. Waggoner v. Midwestern Development, Inc. *North Western Reporter*, 2nd Series, 154: 803–809 (1968)

3. Bradley v. Brucker, *Montgomery County Law Reports*, 69:38, as cited in 1968. "Liability of builder-vendor or other vendor of new dwelling for loss, injury, or damage occasioned by defective condition thereof," *American Law Reports*, 3rd Series, 25:383–441

4. Shirley v. Dracket Products Co., *Product Liability Reports*, para. 6493 (1971)

5. Alfieri v. Cabot Corp., *Product Liability Reports*, para. 5158 (1964)

6. Heritage v. Pioneer Brokerage and Sales, Inc., *Product Liability Reports*, para. 8521 (1979)

7. Harig v. Johns-Manville Products Corp., *Product Liability Reports*, para. 8343 (1979)

8. Wallinger v. Martin Stamping and Stove Co., *Product Liability Reports*, para. 6006 (1968)

9. Dover Corp., and J.R. Preis d/b/a Coastal Bend Sales, v. Perez, *Product Liability Reports*, para. 8581 (1980)

10. Kirsch, Lawrence S., "Behind Closed Doors: The Problem of Indoor Pollutants. Part II," *Environment*, Vol. 25, No. 2 (March 1983)

11. Ibid.

12. Ibid.

13. Ibid.

14. Shabecoff, Philip, "EPA Will Consider Regulation of Formaldehyde," *New York Times* (May 19, 1984), p. 24

15. "High Efficiency Wood Stoves," *Consumer Reports* (Oct. 1985), p. 595

16. Blodgett, Nancy, "The New Outlaws: Smokers Who Light Up at Work," *ABA Journal* (Feb.1, 1986), p. 31

17. Shimp v. Bell Telephone Co., 145 N.J. Super. 516,568 A2d 48 (App. Div. 1976).

18. Lane, Beverly, "Cigarette Maker Breathes Easier," *ABA Journal* (Feb.1, 1986), p. 31

Chapter 6

The Pollution Compendium

1. Wadden, Richard A., and Scheff, Peter A., *Indoor Air Pollution*, p. 33

2. Meyer, Beat, *Indoor Air Quality*, p. 170

3. Meyer, Beat, *Indoor Air Quality*, p. 172

4. Wadden, Richard A., and Scheff, Peter, A., *Indoor Air Pollution*, p. 15

5. National Research Council, *Indoor Pollutants*, pp. VII–50–53

6. National Research Council, *Indoor Pollutants*, pp. VII–21–27

7. Meyer, Beat, *Indoor Air Quality*, p. 278

8. National Research Council, *Indoor Pollutants*, pp. VII–52–55

9. Vedal, Sverre, "Epidemiological Studies of Childhood Illness and Pulmonary Function Association with Gas Stove Use," *Indoor Air and Human Health*, Gammage, R.B. and Kaye, S.V., eds., p. 304

10. National Research Council, *Indoor Pollutants*, p. VII–60

11. Consumer Product Safety Commission, *Status Report on Indoor Air Quality Monitoring Study in 40 Homes*.

12. Sandia National Laboratories, *Indoor Air Quality Handbook—For Designers, Builders, and Users of Energy-Efficient Residences* (Albuquerque, N.M.: U.S. Department of Energy, 1982), p. 67

13. Meyer, Beat, *Indoor Air Quality*, p. 273

14. Sandia National Laboratories, *Indoor Air Quality Handbook—For Designers, Builders, and Users of Energy-Efficient Residences*, p. 53.

15. MacLeod, Kathryn, "Polychlorinated Biphenyls in Indoor Air," *Environmental Science & Technology* (Vol. 15, No. 8, August 1981), pp. 926–928

16. Ibid.

17. The Land and Water Resources Center, University of Maine at Orono, and The Division of Health Engineering, Maine Department of Human Services, *Radon in Water and Air*. (Orono: 1983)

18. National Research Council, *Indoor Pollutants*, p. VII–71

19. Ibid. p. VII–75

20. Ibid.

21. Ibid. p. VII–76

Index